Pioneers in Arts, Humanities, Science, Engineering, Practice

Volume 27

Series Editor

Hans Günter Brauch, Peace Research and European Security Studies
(AFES-PRESS), Mosbach, Germany

More information about this series at http://www.springer.com/series/15230
http://www.afes-press-books.de/html/PAHSEP.htm
http://www.afes-press-books.de/html/PAHSEP_LewisFryRichardson.htm

Nils Petter Gleditsch

Editor

Lewis Fry Richardson: His Intellectual Legacy and Influence in the Social Sciences

Editor
Nils Petter Gleditsch
Peace Research Institute Oslo (PRIO)
Oslo, Norway

Acknowledgement: The photograph on the internal title page was taken from the English version of Wikipedia.en, at: https://en.wikipedia.org/wiki/Lewis_Fry_Richardson.

ISSN 2509-5579   ISSN 2509-5587   (electronic)
Pioneers in Arts, Humanities, Science, Engineering, Practice
ISBN 978-3-030-31588-7   ISBN 978-3-030-31589-4   (eBook)
https://doi.org/10.1007/978-3-030-31589-4

Copyediting: PD Dr. Hans Günter Brauch, AFES-PRESS e.V., Mosbach, Germany

This Springer imprint is published by the registered company Springer Nature Switzerland AG
The registered company address is: Gewerbestrasse 11, 6330 Cham, Switzerland

The original version of this book was revised: Reference author name "Gregory D. Hess" have been replaced with George D. Hess in page numbers 70, 83, 98, 111, and 145. The correction to this book can be found at https://doi.org/10.1007/978-3-030-31589-4_12

Contents

Chapter 1
Lewis Fry Richardson – A Pioneer Not Forgotten

Nils Petter Gleditsch

Abstract Lewis F Richardson, a physicist by training, remains a towering presence in two academic subjects, meteorology and peace research. Prizes are named for him in both fields. This chapter introduces a collection of articles assessing Richardson's legacy and his enduring influence in the social sciences. It reviews his citations as an indication of the range of his influence and discusses his impact in five areas of social science: the study of arms races, data collection on deadly quarrels, the stability of the long peace, the role of geography in conflict, and the role of mathematics in peace studies. It also includes a brief discussion of the conscience of a scholar with regard to preparations for war.

1.1 His Life and Work

Lewis F Richardson was trained as a physicist, but gained his fame first in meteorology and then in the study of conflict. Although he never gained employment at a leading university, his work in meteorology was widely respected by his contemporaries and has remained among the foundations of the field. His work on conflict was seen as more unorthodox. Certainly, his formal models and quantitative empirics were well ahead of the curve in the discipline of international relations in his lifetime. It was not until seven years after his death that his two major volumes on conflict found a publisher (Richardson, 1960a, b).

Since then, Richardson has been honored in various ways. In 1972, British Prime Minister Edward Heath opened a new wing of the Headquarters Building of the

My work on this volume was supported by the Gløbius fund and by the Conflict trends project (#402561). Most of the chapters were first presented to two sessions on Richardson at the 59th Annual Convention of the International Studies Association, San Francisco, 4–7 April 2018. I am grateful to Håvard Hegre, Gerald Schneider, and the contributors to this volume for comments on my introduction.

N. P. Gleditsch (ed.), *Lewis Fry Richardson: His Intellectual Legacy and Influence in the Social Sciences*, Pioneers in Arts, Humanities, Science, Engineering, Practice 27, https://doi.org/10.1007/978-3-030-31589-4_1

Meteorological Office named the Richardson Wing.[1] The Department of Mathematics at the University of York has sponsored a Lewis F Richardson lecture series since 2015.[2] Unusually, scientific prizes are named for him in both his main fields. In 1960, the Royal Meteorological Society established the annual LF Richardson Prize for meritorious papers by young authors in one of the journals of the society.[3] Since 1997, The European Geosciences Union has awarded the Lewis Fry Richardson Medal for 'exceptional contributions to nonlinear geophysics in general'.[4] And from 2001, scholars who have spent most of their academic life in Europe and who have made exemplary scholarly contributions to the scientific study of militarized conflict, have been honored with the Lewis F Richardson Lifetime Award, with Michael Nicholson as the first recipient.[5] As I have experienced on a couple of occasions, if a conflict researcher gets an opportunity to speak to a group of meteorologists (say on the topic of climate change and conflict), a favorable mood can be generated by an early reference to Richardson.

Richardson was in many ways a loner. Although he carried out an extensive correspondence and was receptive to criticism of his own work – in fact, his two major volumes contain a number of fictional dialogues with his critics – he generally worked without assistants, and most of his work is single-authored. He often worked under difficult conditions. The extreme case is his work on meteorology while serving as an ambulance driver in France in World War I. In 1917, during the battle of Champagne, he sent his working copy of the manuscript on weather prediction 'to the rear, where it became lost, only to be re-discovered some months later, under a heap of coal' (Richardson, 1922: ix). Of course, as befitting a scholar of his generation, he relied very heavily on his wife Dorothy not just for moral support but in the practical work of carrying out experiments and in copy-editing.[6]

As is evident from the timeline in the Appendix, Richardson spent most of his professional life in positions where he either worked on practical problems or taught science at the basic level, notably at Paisley Technical College (1929–40). Apparently, Richardson was not the world's best teacher, but he is described as 'conscientious and caring' (Ashford, 1985: 150f). Much of his research was carried out in his spare time. It was only after retirement, for the last 13 years of his life, that he was able to devote himself full-time to research.

Richardson's publications in meteorology, notably *Weather Prediction by Numerical Process* (Richardson, 1922) and a later article on atmospheric diffusion (Richardson, 1926), remain his most frequently cited items. The 1926 article is recorded with well over 1000 citations on Web of Science, including 42 citations in

---

[1] Ashford (1985: 246ff).

[2] www.york.ac.uk/maths/events/lfr/.

[3] Ashford (1985: 245), www.rmets.org/our-activities/awards/l-f-richardson-prize.

[4] www.egu.eu/awards-medals/lewis-fry-richardson/.

[5] http://ksgleditsch.com/richardson_award.html.

[6] Ashford (1985: 239f).

the first seven months of 2018![7] Among his social science writings, his two posthumously published books top the list, with *Arms and Insecurity* (Richardson, 1960a) a little ahead of *Statistics of Deadly Quarrels* (1960b). Both of these books continue to be cited to this day, although not at the level of his 1926 article. Richardson (1961), another posthumous publication, is also widely cited.

For several years, Richardson maintained a strong interest in psychology, and delved into topics like intelligence, the quantitative assessment of pain, perception, and national hatred. He published in scholarly journals, including several articles in *British Journal of Psychology*, attended professional meetings, and even went to the trouble of acquiring an academic degree in psychology at the age of 48.[8] He also taught a psychology course in college. Several of Richardson's articles in psychology are respectably cited, particularly his work on the measurement of sensations (Poulton, 1993). But on the whole he appears to have had more limited impact in this field, although some of his methods have been widely adopted. Richardson's first major publication on conflict was, characteristically, titled *The Mathematical Psychology of War* (Richardson, 1919). That he focused on psychology rather than war when he more or less left meteorology in the 1920s, has been explained as a result of a hope that World War I had been so devastating that another major war seemed unlikely. When political and military developments turned to the worse in the 1930s, Richardson devoted almost all his research time to the question of war and peace (Nicholson, 1999: 543).

1.2 The Study of Arms Races

The notion of an arms race is an old one and extends well beyond the field of international relations, e.g. in biological studies of adaptation and counter-adaptation between predators and prey (Smith, 2020, in this volume: 8–9). Boulding (1962: 25) suggested the label 'Richardson processes', since he had provided the most extensive theoretical treatment. Richardson was concerned with how the acquisition of arms by two or more hostile powers could lead to a competitive race (Richardson, 1960a). He analyzed under what conditions such a race would become unstable and was likely to end in war. He studied a number of arms races from this perspective, notably the arms races preceding the two World Wars. He developed a formal model which was driven by competition with the other side, a 'fatigue' factor determined by the level of one's own military spending, and the 'grievance' against the other side. A very large number of scholars have tried to improve on this model and the empirical measures used to test it. Two major debates have emerged out of this literature: First, whether arms acquisitions are driven mainly by competition, or by internal processes as

---

[7]All searches on ISI Web of Science, 8 August 2018.
[8]Ashford (1985: 112f).

argued, for instance, by Senghaas (1990). Secondly, to what extent arms races are associated with the outbreak of war, as maintained by Wallace (1979) and others.

Both of these debates became quite heated and acquired ideological overtones. Smith (2020) concludes that the lack of specificity in the Richardson model was a strength as well as a weakness. It could be applied in a wide variety of contexts, but at the same time it was difficult to evaluate empirically. Its simplicity also makes it a great teaching tool. Diehl (2020, in this volume) shows how the empirical testing of the arms-race-to-war linkage has become more sophisticated since Richardson's original analyses, notably in examining the no-arms race cases. No clear consensus has emerged, but Richardson continues to provide an inspiration to study the role of arms races in raising the risk of war.

The bulk of the work on arms races has focused on competition between two hostile powers. However, Richardson also developed extensions of his arms race model to three or more nations (Richardson, 1960a, Chs 15, 17). The greater complexity of these models makes it harder to derive the conditions for stability, and the empirical testing also becomes much more complicated. However, as argued by Michael Ward (2020, in this volume), Richardson's work points the way towards a network perspective on international affairs. And recent progress in data collection and processing makes it much more feasible to simulate complicated systems of interdependent processes.

1.3   Identifying Deadly Quarrels

Unlike present-day scholars, Richardson could not pick a suitable dataset off the shelf in order to test his theories of the causes of war. He had to develop his own, and Richardson (1960b) is the final result of this effort, listing all 'fatal quarrels' after 1820 which had ended. He consulted a large number of historical sources as well as police statistics and the notes on each conflict list the sources used. In parallel endeavors, Sorokin (1937) and Wright (1942) also developed datasets on wars. Richardson only became aware of these lists when his own was largely complete, and in his book he comments on some similarities and differences. In his own list, inclusion was determined strictly by the number of deaths, which he believed to 'the most reliable method for statistical purposes' (Richardson, 1960b: 5). A 'deadly quarrel' is defined as 'any quarrel which caused death to humans (p. 6). This includes not just wars, but also 'murders, banditries, mutinies, insurrections', but not 'accidents, and calamities such as earthquakes and tornadoes' or indirect deaths from famine and disease.

The two most frequently used datasets in current empirical research on armed conflict, the Correlates of War (COW) Project[9] and the Uppsala Conflict Data

---

[9]http://www.correlatesofwar.org/.

Program (UCDP),[10] draw directly or indirectly on the earlier lists, but have settled for narrower definitions than Richardson. For instance, UCDP defines an armed conflict as 'a contested incompatibility that concerns government and/or territory where the use of armed force between two parties, of which at least one is the government of a state, results in at least 25 battle-related deaths in one calendar year'.[11]

The trend in recent empirical work on armed conflict has been in the direction of greater inclusion, although not necessarily by including everything in a master category such as 'deadly quarrels'. The COW project's first dataset (Singer & Small, 1972) included interstate wars only, while the current lists also include data on civil wars, extra-state wars, and non-state wars. While UCDP started out with data on three types of conflicts – interstate, intrastate, and extra-state – it has later added data on non-state conflicts and on one-sided violence. These data are reported in separate spreadsheets, but it is possible to merge the information, and UCDP now reports an annual world total for deaths in all these forms of violence (Pettersson & Eck, 2018).

There is some controversy over the issue whether or not to include deaths from crime in the study of armed conflict. The study of violent crime is usually conducted quite separately from the study of war. But the basic framework, where actions are seen as a function of motive and opportunity, is the same, and the recent decline in homicide rates is frequently interpreted as part of the same turn away from violence as the decline in war casualties (Pinker, 2018). In a controversial study, the International Institute of Strategic Studies reported in 2017 that Mexico had the world's second-most-lethal conflict in 2016 (after Syria) (IISS, 2017: 5).[12] But this assumed that all or most of the murders in Mexico (23,000) were connected to organized crime, whereas UCDP limits non-state conflict to the use of armed force between two organized groups and attributed a much smaller number (1300) of deaths in Mexico to this type of armed conflict. Since deaths from homicides vastly exceed casualties in war or civil war in most countries and most years, it matters a great deal how large a share is classified as armed conflict. It is essential that the classification criteria are the same across time and space.

Richardson reported all casualties in his deadly quarrels in logarithms to the base ten. He was wary of fictitious accuracy. He commented on three seemingly very different estimates of the number of deaths in the Union army in the American Civil War (359,528, 279,376 and 166,623). Reporting the logarithms (5.6, 5.4, and 5.2) 'brings out their substantial agreement' (Richardson, 1960b: 7). The rationale was the same as for measuring the severity of earthquakes with the Richter scale (Nicholson, 1999: 550) or its successor the Moment Magnitude Scale.

---

[10]http://ucdp.uu.se/.

[11]www.pcr.uu.se/research/ucdp/definitions/.

[12]For critiques of the IISS report, see e.g. Phillips (2017) and Estévez-Soto (2017).

1.4   The Stability of the Long Peace

A number of scholars have argued that the world is becoming less violent if the violence is measured by casualties in armed conflict (Lacina, Gleditsch & Russett, 2006; Pinker, 2018). Others question the stability of this trend and outline plausible scenarios that could produce a large war (e.g. Ellsberg, 2017). Richardson (1948, 1960b: Chs 3–4) found that war sizes followed a power-law distribution where the frequency of wars of size x is proportional to $x^{-\alpha}$, where α is a constant. Bigger wars are less common than smaller ones and the value of α determines the rate at which war frequencies decrease as war sizes increase. Aaron Clauset (2020, in this volume) confirms this, using more sophisticated statistical tools and better data than were available to Richardson. His analysis is consistent with a constant hazard of interstate war. This does not contradict the empirical fact of a decline in the lethality of war over the last 70 years, but Clauset concludes that the pattern of relative peace would have to last another 100 years before one can conclude that it is a statistically significant trend rather than the result of chance. Michael Spagat & Stijn van Weezel (2020, in this volume) do not dispute this. However, they point out that when the fatalities are measured relative to the size of the population, measuring the risk that a random person will suffer a battle-related death, the evidence for a real change becomes stronger. The same is true if one moves the hypothesized break-point forward to 1950 rather than the end of World War II. Finally, when including civil wars, the no-change hypothesis can be rejected with confidence. This debate will no doubt continue to inspire a host of new studies. Richardson is frequently quoted for his statement that his 'equations are merely a description of what people would do if they did not stop to think' (Richardson, 1960a: 12). Indeed, Clauset (2020: 124–125) finds it puzzling that the hazard of war should remain constant, given the non-stationarity of human population, the number of recognized states, commerce, communication etc. If the straitjacket of the power-law distribution for the size of wars can be broken, Richardson would probably have been delighted to find that people had actually stopped to think.

1.5   The Role of Geography in Conflict

Meteorology is very much a geographical science. Weather prediction depends crucially on estimates of how and when air pressures and sources of precipitation move geographically. It is not surprising when Richardson took his tools from physics and set them to work on conflict, that geographical considerations should permeate his writings in the new field.

Richardson was one of the first to write about the importance of contiguity to fighting. 'The obvious reason why the murderer and his victim were usually subjects of a common government is their localization.' (Richardson, 1960b: 297). He therefore needed to develop appropriate measures for geographical opportunities for fighting. One of these was the length of a common border between two countries. Here, Richardson was the first to point out that this length depended on the scale of measurement. The shorter the yardstick, the longer the boundary (Richardson, 1961/1993: 607ff). Although ignored at the time, this would later inspire Mandelbrot's work on fractals (Mandelbrot, 1967).

As noted by Gleditsch & Weidmann (2020) and Scheffran (2020), both in this volume, Richardson pioneered the use of cell-based approaches to conflict analysis, long before the introduction of Geographic Information Systems (GIS). He noted that the number of 'conceivable belligerents' in a civil war could not simply be determined by what groups had actually fought. Such potential parties to fighting could not be identified from works of history 'because insurgents were often not recognized as a group until they had declared themselves to be such by revolting.' (Richardson, 1960b: 307). Richardson therefore estimated the number of cells of equal numbers of people (potential conflict actors) and discussed how 'local pacifying influences', such as common government, language, or religion, might reduce the risk of civil war between them. The rapid growth of GIS and of spatial datasets on political, demographic, socioeconomic, and environmental character- istics of subsets of nations, has led to a major reorientation of empirical analyses of war and peace, as noted by Gleditsch & Weidmann (2020).

Richardson also pointed out a curious fact about national boundaries: There are no examples of four countries meeting in a single point, as in the Four Corners area of the US. He attributed this to the role of warfare in shaping boundaries (Gleditsch & Weidmann, 2020: 73f).

Scheffran (2020) suggests that Richardson's conflict model offers a basis for insights in the potential impacts of climate change on conflict and cooperation. This is a credible extension of his model. Given Richardson's enduring interest in the weather and in conflict, it is a fascinating thought that he might have been a pioneer in the now blooming research area on climate, weather, and conflict (Buhaug, 2016). However, while his *Statistics of Deadly Quarrels* has chapters on several potential causes of war (such as poverty, language, religion, and contiguity), there is no chapter on climate or the weather. Indeed, the only place I have been able to find a link of sorts is on p. 129 where he cites an observation by Quincy Wright that wars in the north temperate zone have ordinarily begun in spring or summer.

1.6 The Role of Mathematics in Peace Studies

The discipline of international relations in Richardson's time, including the study of war and peace, was case-oriented and strongly influenced by legal and normative considerations. Attempts at generalizations were rarely based on systematic data. Along with Pitirim Sorokin and Quincy Wright, Richardson was one of the few pioneers in what today is a vibrant field of quantitative conflict studies.

Richardson's achievements in this regard is all the more remarkable in that he seemed largely unaware of the introduction of mathematical models in other social sciences, particularly in economics, which was taking place at the same time (Nicholson, 1999: 556).

Much current work in international relations is cross-sectional because temporal data are lacking. Kelly Kadera, Mark Crescenzi & Dina Zinnes (2020, in this volume) point out that in his work on the dynamics of conflict, Richardson was a pioneer in investigating the role of time in international relations. They argue that studies in the Richardson tradition using differential equations model time more explicitly than most game theory models, which focus on equilibria. As Nicholson (1999: 547) points out, most economists of his era would probably have approached the arms race as a problem in comparative statics, 'where determining the equilibrium was the main problem and the paths along which the system moved to achieve it was a subsidiary issue if considered at all.'

While early empirical analyses of the conditions of war and peace tended to look at the influence of one variable at a time, Richardson was clearly sensitive to multivariate analysis with interactions between factors: As Smith (2020: 26) notes, he recognized that a common border would increase the probability of war, but also amount of trade, which may in turn have a pacifying effect.' This basic point was frequently overlooked in many early studies of the trade-conflict relationship (Schneider, Barbieri & Gleditsch, 2003).

Niall MacKay (2020, in this volume) offers a comparison between Lanchester's model of war attrition and the Richardson arms race model. Their starting-points were quite different. Richardson was concerned with the hazard of war and how arms races could be prevented or limited. Lanchester was interested in how to win a war. Both are models of two-way interaction. Both can be generalized from duels to 'truels'. In both models, a scholar can work out the conditions for a stalemate. MacKay discusses the possibilities of combining insights from the two models.

1.7   The Conscience of a Scholar

Richardson came from a Quaker background and his religious affiliation had a pervasive influence on his life and career. His interest in psychology was apparently inspired by the social service of the Society of Friends (Ashford, 1985: 51). When World War I broke out, he was working for the Meteorological Office in Eskdalemuir. Given the national importance of his work there, he could probably have continued until the end of the war. Instead, he applied for leave from the Met Office to join the Friends Ambulance Unit in France as a conscientious objector. His application for leave was turned down, and he eventually resigned from his position in order to join the ambulance unit in 1916. He served there for nearly three years. While in France, he wrote some of his early papers on war and peace. He rejoined the Met Office in 1919 at Benson Observatory. However, in 1920 the Met Office was transferred to the Air Ministry, and Richardson resigned. As he wrote to the Norwegian meteorologist Vilhelm Bjerknes, 'I do not like preparations for war' (Ashford, 1985: 105). Richardson nevertheless continued to publish papers in meteorology in the 1920s. In the 1930s, researchers in chemical warfare became interested in Richardson's work on atmospheric turbulence and made 'delicate approaches', causing 'a time of heart-break' according to his wife. He then destroyed research results that had not been published (Körner, 1996: 189).

Not all Quakers or other British pacifists reacted to World War I in the same way that Richardson did. For some, national patriotism trumped their peaceful principles. This was, of course, even more so in World War II, because the enemy seemed particularly evil. Körner (1996: 207), in a sympathetic review of Richardson's work, is left uneasy by Richardson's refusal to assign blame for any conflict, although he acknowledges that participants in a deadly conflict routinely accuse the other side of starting it.

Scholars today are faced with similar dilemmas. Is it immoral to do research on weapons of mass destruction? Or is it immoral to leave the field to the other side? Traditionally, such dilemmas have been faced mainly by natural scientists. However, the military and intelligence services are also increasingly interested in the social sciences. In the US, for instance, they are heavily involved in funding the social sciences through the Minerva Research Initiative (https://minerva.defense.gov/) and the Political Instability Task Force.[13] The research is unclassified and published openly and has spawned a number of seminal books and articles. The policy orientation of these funding initiatives is not in doubt and they serve as vehicles for bringing social scientists and policymakers in closer touch. Critics might argue, as they did in the mid-1960s when the US Department of Defense started recruiting social scientists for a counterinsurgency program called Project Camelot (Horowitz, 1967), that such efforts aid US policymakers in cementing a hierarchical and unjust international order. Others would respond that it would be counterproductive to leave major sources of social science funding to ideologues or less competent social scientists. There are no easy answers to these dilemmas, but those who shy away from research sponsorship under the rubric of national security today, probably sacrifice less in career terms than did Richardson.[14]

Despite Richardson's devotion to theory, he was not an impractical scientist unconcerned with practical implementation. He developed ingenious procedures for carrying out experiments and held several patents. He wrote two papers on voting procedures in international organization and seemed to have been convinced that if he could persuade decision makers of the hazard of arms races and war preparations, he could help to prevent them. It was precisely because he was concerned

---

[13]PITF does not appear to have an official website, but a Wikipedia article explains the history and nature of the research sponsorship: https://en.wikipedia.org/wiki/Political_Instability_Task_Force.

[14]There is little public discussion of these issues, but one member of the PITF resigned after the election of Donald Trump, with a harsh indictment of his colleagues ('academic courtiers') who preferred to remain silent in the face of a situation where 'the greatest source of political instability in the world will be the administration of the US Federal Government.' Cf https://scatter.wordpress.com/2017/01/20/why-i-resigned-from-the-political-instability-task-force/.

with putting science into practice, that he was wary of contributing to the preparation for war. One of the things he had learned from a teacher at Bootham School in York, was that 'science ought to be subordinate to morals.' (Gold, 1954: 218)

1.8   The Impossible Dream

Richardson helped to create a new field of research virtually from nothing. In this brief introduction, I have focused on some of the key areas of his research on war and peace, but those who consult *Statistics of Deadly Quarrels* will find chapters on languages and war, religions and war, economic causes of war, and many others. One of his conclusions on interreligious wars speaks directly to an important current debate: 'There were more wars between Christians and Moslems than would be expected from their populations, if religious differences had not tended to instigate quarrels between them.' (Richardson, 1960b: 245)

Richardson was ahead of his times in approaching the question of war and peace with tools he had acquired in his work in physics. He was unafraid to tackle problems that were hard to solve, or even insoluble with the resources available to him at the time. His work on weather forecasts required an enormous number of calculations just to predict tomorrow's weather from today's. In fact, his own attempts at weather prediction took longer than the passage of the actual weather. Unfazed, Richardson calculated that a 'staggering' staff of 64,000 (human) computers would be needed to complete a weather forecast before the deadline. (Actually, there was an error in his calculations, the correct figure was 256,000.) 'Perhaps in some years' time it may be possible to report a simplification of the process.'[15] Indeed! With the advent of digital computers, predicting the weather using the methods introduced by Richardson has become standard practice.

With the benefit of hindsight, we can see that his work was given less attention at the time of publication than it deserved. Cambridge University Press agreed to publish his now celebrated book *Weather Prediction by Numerical Process* in 1922 only after receiving subsidies from the Royal Society and the Met Office. It was printed in just 750 copies and sold even fewer. This and similar experiences have led to Richardson being portrayed as neglected genius. Nevertheless, his work did in fact inspire early pioneers in peace research like Kenneth Boulding, Karl Deutsch, Anatol Rapoport, J David Singer, and Quincy Wright (Nicholson, 1999: 555, 559). The very first issue of *Journal of Peace Research* contained an article on Richardson's arms race model (Smoker, 1964) and the first volume of *Journal of Conflict Resolution* a whole special issue (1957, 3) on Richardson. Dina Zinnes (2020, in this volume) explains how her encounter with Richardson came to determine the direction of her own distinguished career in the field. No doubt, the work of Lewis Fry Richardson will continue to inspire new generations of scholars.

---

[15]Richardson (1922: 219). Cf Ashford (1985: 91f).

References

Chapters in this volume have been omitted in the reference list.

Ashford, Oliver M (1985) *Prophet – or Professor? The Life and Work of Lewis F Richardson.* Bristol: Hilger.

Ashford, Oliver M et al. (eds) (1993) *Collected Papers of Lewis F Richardson,* 2: Quantitative Psychology and Studies of Conflict. Cambridge: Cambridge University Press.

Boulding, Kenneth E (1962) *Conflict and Defense. A General Theory.* New York: Harper and Brothers.

Buhaug, Halvard (2016) Climate change and conflict: Taking stock. *Peace Economics, Peace Science and Public Policy* 22(4): 331–338.

Clauset, Aaron (2020) On the frequency and severity of interstate wars. Ch. 10 in this volume.

Diehl, Paul F (2020) What Richardson got right (and wrong) about arms races and war. Ch. 4 inthis volume.

Ellsberg, Daniel (2017) *The Doomsday Machine: Confessions of a Nuclear War Planner.* London: Bloomsbury.

Estévez-Soto, Patricio R (2017) A report says that Mexico is the second-deadliest conflict zone in the world –it's just not true. *The Conversation,* 18 May, http://theconversation.com/a-report-says-that-mexico-is-thesecond-deadliest-conflict-zone-in-the-world-its-just-not-true-77898.

Gleditsch, Kristian Skrede & Nils B Weidmann (2020) From hand-counting to GIS: Richardson inthe information age, Ch. 7 in this volume.

Gold, Ernest (1954) Lewis Fry Richardson. *Obituary Notices of Fellows of the Royal Society* 9: 217–235, https://royalsocietypublishing.org/doi/10.1098/rsbm.1954.0015.

Horowitz, Irving Louis (ed.) (1967) *Rise and Fall of Project Camelot.* Cambridge, MA: MIT Press.

IISS (2017) *Armed Conflict Survey 2017.* London: International Institute of Strategic Studies.

Körner, Thomas William (1996) Ch. 8 A Quaker mathematician and Ch. 9 Richardson on War. In: *The Pleasures of Counting.* Cambridge: Cambridge University Press, 159–193, 194–227.

Lacina, Bethany; Nils Petter Gleditsch & Bruce Russett (2006) The declining risk of death in battle. *International Studies Quarterly* 50(3): 673–680.

MacKay, Niall (2020) When Lanchester met Richardson: The interaction of warfare withpsychology. Ch. 9 in this volume.

Mandelbrot, Benoît (1967) How long is the coast of Britain? Statistical self-similarity and fractional dimension. *Science* 156(3775): 636–638.

Nicholson, Michael (1999) Lewis Fry Richardson and the study of the causes of war. *British Journal of Political Science* 29(3): 541–563.

Pettersson, Thérése & Kristine Eck (2018) Organized violence, 1989–2017. *Journal of Peace Research* 55(4): 535–547.

Phillips, Brian J (2017). Is Mexico the second-deadliest 'conflict zone' in the world? Probably not. MonkeyCage, *Washington Post,* 18 May, www.washingtonpost.com/news/monkey-cage/wp/2017/05/18/is-mexico-thesecond-deadliest-conflict-zone-in-the-world-probably-not/.

Pinker, Steven (2018) *Enlightenment Now: The Case for Reason, Science, Humanism and Progress.* New York: Penguin.

Poulton, EC (1993) The quantifying of mental events and sensations. In: Ashford et al., 491–514.

Richardson, Lewis F (1919) *The Mathematical Psychology of War.* London: Hunt.

Richardson, Lewis F (1922) *Weather Prediction by Numerical Process.* Cambridge: Cambridge University Press.

Richardson, Lewis F (1926) Atmospheric diffusion shown on a distance-neighbour graph. *Proceedings of the Royal Society A* 110(756): 709–737.

Richardson, Lewis F (1948) Variation of the frequency of fatal quarrels with magnitude. *Journal of the American Statistical Association* 43(244): 523–546. Reprinted in Ashford et al., 577–627.

Richardson, Lewis F (1960a) *Arms and Insecurity: A Mathematical Study of the Causes and Origins of War.* Pittsburgh, PA: Boxwood.

Richardson. Lewis F (1960b) *Statistics of Deadly Quarrels*. Pittsburgh, PA: Boxwood.

Richardson, Lewis F (1961) The problem of contiguity: An appendix of Statistics of Deadly Quarrels. *General Systems Yearbook* 6: 139–186. Reprinted in Ashford et al., 579–627.

Schneider, Gerald; Katherine Barbieri & Nils Petter Gleditsch (eds) (2003) *Globalization and Armed Conflict*. Lanham, MD: Rowman & Littlefield.

Senghaas, Dieter (1990) Arms race dynamics and arms control. Ch. 2 in: Nils Petter Gleditsch & Olav Njølstad (eds) *Arms Races: Technological and Political Dynamics*. London: Sage, 15–30.

Singer, J David & Melvin Small (1972) *The Wages of War, 1816–1965: A Statistical Handbook*. New York: Wiley.

Smith, Ron P (2020) The influence of the Richardson arms race model. Ch. 3 in this volume.

Smoker, Paul (1964) Fear in the arms race. *Journal of Peace Research* 1(1): 55–64.

Sorokin, Pitirim (1937) *Social and Cultural Dynamics: Fluctuations of Social Relationships, War, and Revolution*. New York: American Book Company.

Spagat, Michael & Stijn van Weezel (2020) The decline of war since 1950: New evidence.

Wallace, Michael D (1979) Arms races and escalation: Some new evidence. *Journal of Conflict Resolution* 23(1): 3–16.

Wright, Quincy (1942) *A Study of War*. Two volumes. Chicago, IL: University of Chicago Press. Updated version in one volume, 1965.

Nils Petter Gleditsch, b. 1942, is Research Professor at the Peace Research Institute Oslo and Professor Emeritus of Political Science at NTNU. He served as Director of PRIO (1972, 1978–79) and as Editor of *Journal of Peace Research* (1977, 1983–2010) and was President of the International Studies Association in 2008–09, nilspg@prio.org.

Chapter 2
Lewis Fry Richardson: A Personal Narrative

Dina A. Zinnes

Abstract This chapter is the personal story of how the author, just out of graduate school, encountered Richardson's two posthumous volumes, *Arms and Insecurity* and *Statistics of Deadly Quarrels*, and how these volumes helped her resolve key issues that had troubled her. What is theory and how does it differ from a model? The first volume pushed her to learn mathematics and, through various twists and turns, eventually to an understanding of the power of a mathematically written story. The second volume provided insights into data collection and measurement as well as a deeper understanding of the mechanics of mathematical modeling. Thus, the two volumes together gave her the basis for finally answering the questions from her graduate school days.

2.1 Introduction

When I finished graduate school in political science, I was left with a set of troubling, unanswered questions. I had found the debates of realism and idealism unsatisfying and believed that a science of international politics was both needed and possible. But I was baffled and confused about what I believed were the critical pieces of a science. What is 'theory'? Why is it important? What is a 'model' and how does it differ from 'theory'? Where do mathematics, statistics, and data fit? It was the work of Lewis Fry Richardson that, over the years, pushed and nudged me along a journey that finally led me to the answers. The combined volumes of *Arms and Insecurity* Richardson (1960a) and *Statistics of Deadly Quarrels* Richardson (1960b) became the 'intellectual bibles' for my search. They were bewildering at times, frustratingly challenging at others, but they ultimately teased me into an understanding of the bits and pieces that transform a study of international politics into a science of international politics.

This is a revised version of a paper presented at the Richardson panel at the 59th Annual Convention of the International Studies Association, San Francisco, CA, 7 April 2018. I am grateful to participants in the panel for comments.

13

N. P. Gleditsch (ed.), *Lewis Fry Richardson: His Intellectual Legacy and Influence in the Social Sciences*, Pioneers in Arts, Humanities, Science, Engineering, Practice 27, https://doi.org/10.1007/978-3-030-31589-4_2

My journey with Richardson began when I was a graduate student at the University of Michigan. *Journal of Conflict Resolution* had recently been founded and the graduate students were made part of the organization team. An initial topic of conversation was how and when a special issue might be devoted to this meteorologist's unpublished notes on conflict and war. Work was under way to compile the extensive notes into publishable book form following Richardson's death, but it was felt that the significance of the material might go unrecognized by the potentially most relevant audience – the social sciences – because of its mathematical nature. Thus, it seemed an obvious idea to combine the launch of a journal dedicated to analytical research on conflict with a layman's introduction to Richardson's mathematical approach to issues of conflict. A long review article by Rapoport (1957) in a special issue of the journal served as my introduction to this amazing work.

When *Arms and Insecurity* was published in 1960, I was eager to tackle the 'real thing'. But while the book was captivating, it also posed enormous challenges. The 'dialogues' – Richardson talking to himself as he puzzled his way through a question – were exciting and provided a wonderful insight into the strategy of thinking about a research problem. When the dialogues led to the 'story' of two statesmen arguing how best to protect their country against the other, I was intrigued. But when the story was translated into two equations, confusion began. Richardson waived a mathematical wand, and out of the equations came the 'explosion' of 'war' – and confusion turned into amazement. It was magic. Richardson had pulled a rabbit out of a hat!

Since I had not taken mathematics in college, my mathematical training had ended with high school trigonometry. I understood algebraic equations, but I had never encountered the symbol dx/dt. Moreover, Richardson's ability to move from verbal language to mathematics was baffling. My only encounters with the transition between verbal language and mathematics were word problems: How long would it take to go from X to Y when traveling at …? But most mystifying of all was Richardson's ability to draw conclusions about 'war' and 'peace'. How did he know that certain inequalities between parameters of the equations would produce an 'explosion?' Where did these rabbits come from?

Puzzling over the equations for hours forced me to conclude that my mathematical training was deficient. If I was going to be able to understand Richardson's work, I had to go back to school. I had recently finished my Ph.D. and followed my husband to his first academic job at Indiana University, so there was a breathing spell of free time to explore mathematics. I began with the basics – calculus – and proceeded through the 'bread and butter' sequence of college mathematics courses. It was a slow and, at many points, difficult process – mathematics was a very different world for someone coming from the social sciences. But bit by bit I acquired the basic vocabulary, grammar, and mode of thinking of this new and intriguing world.

When my first teaching job materialized at Indiana University, I came across a book that seemed to compliment Richardson's work. *Introduction to Models in the Social Sciences* by Lave & March (1975) was similar to Richardson in three ways. First, Richardson had begun his arms race model with a question: Why do two peaceful states end up in a violent conflict? Similarly, Lave & March illustrate the importance of starting with a question. They explore a series of simpler questions: Why do high school friends get assigned to adjacent rooms in freshman college dorms? Why are the students who ask silly questions in class typically athletes? Why do women who attend all-women high schools perform at a higher level in college than women at co-educational high schools?

Second, Richardson's question led him to formulate a story about two statesmen from two different states wanting to protect their respective countries from possible aggression by the other. Similarly, Lave & March provide verbal stories that answer each of the questions raised in their examples. Third, both Richardson and Lave & March draw conclusions from their stories. In each case the authors use their stories to predict something new, an observable phenomenon that, as Lave & March put it, would follow if the stories were in fact 'true'.

But there was a significant difference between the authors. Unlike Richardson, Lave & March never translate their stories into mathematics. Instead, they draw verbal conclusions from each story. As I examined these verbal conclusions, I found myself troubled. Some seemed direct and reasonable, but others were far less clear. This was an intriguing contrast with Richardson's analyses. Although I was still not competent to follow all the details, Richardson's conclusions appeared less ambiguous and far more convincing. Thus, Lave & March (1975) reinforced the significance of questions and stories, but simultaneously made me aware of the potential value of translating a story into mathematics.

As I moved further into my research career the focus on theory and mathematical modeling was overshadowed by another part of the science puzzle. Singer (1969) argued that if the study of international politics was to go beyond the philosophical realist/idealist debates to become a science, it was critical that arguments be subjected to empirical verification. Singer called for brush-clearing research in which the major hypotheses of the field would be evaluated with data.

Obtaining data relevant to international issues, however, was not a simple matter. Thus, the focus of the newly emerging field of 'quantitative international politics' turned to issues concerning data: How do you define the variables of interest (e.g. national power, conflict, war, crisis), what are the appropriate sources from which to extract 'data', how can data extraction be made reliable? Some argued that relevant data could be created in laboratories as in the Inter Nation Simulation (Guetzkow, 1963) where people played the roles of statesmen and interacted according to rules believed to govern the international system. But the validity issues inherent in these laboratory representations seemed insurmountable and the field moved instead towards real time data and the subsequent creation of large datasets.

The resultant data movement had three prongs: (1) the collection of daily events in contemporary time to better understand crises (e.g. WEIS,[1] COPDAB[2]), (2) the compilation of major historical international crises, e.g. wars (COW[3]), and (3) the collection of national attributes (such as DON[4]). The datasets were typically the work of independent researchers, each concerned with specific questions that dictated both the definition of variables and the type of data to be collected. Consequently, as the various datasets were compiled, comparisons became important and questions arose as to the 'true' meaning of concepts like 'war'. Did one 'operationally' define a 'war' as an overt declaration by one state against another? Or was it an event in which at least 100 combatants were killed? Or was 'war' just the end point of a scale that began with one human killing another?

Fascinated by these issues and wanting to understand the art and science of data collection better, I purchased *Statistics of Deadly Quarrels*, which contained one of the early datasets relevant to conflict and war. Initially, my interest was in the dataset that comprised the first half of the volume. As had been true in *Arms and Insecurity*, the volume began by describing the author's thinking process behind the collection of data and was both captivating and enlightening. Part of the discussion consisted of his definition of the key variable of interest – a 'deadly quarrel'. Richardson's Quaker background made him primarily interested in those situations in which one human killed another, i.e. human conflicts that ended in at least one death. Murder was simply one end of a scale that ended in a world war.

This different but creative way of conceptualizing and measuring a variable made me aware of the fact that 'data' do not exist independent of their operational measurement. The operational definition of a variable would necessarily determine the kinds of questions that could be answered using a given dataset. This realization led me to conclude that collecting data to test a specific hypothesis was important, but collecting data for a data bank could be of limited value. The operational definition used to collect the data for a data bank is unlikely to fit many research questions. In fact, the existence of data banks may have the unfortunate effect of pushing scholars to adjust their research questions to fit the definitions used to collect the data bank variables.

Although my purchase of *Statistics of Deadly Quarrels* had been motivated by Richardson's dataset, I discovered that the book also contained valuable information relevant to my original concerns. It is actually two books in one. While the first part consists of Richardson's discussion of his data collection procedures and the resultant dataset, the second part is, in many respects, an answer to Singer's call for a brush-clearing of old arguments about conflict and war. The second half of the book puts the dataset of the first half up against a variety of age-old arguments: Do borders cause wars? Do differences in religion lead to conflict? Is economics the

---

[1]McClelland (1978).
[2]Azar (2009).
[3]Singer & Small (1972).
[4]Rummel (1972).

source of conflict? Each chapter looks at one of these questions and, using the dataset, either formulates hypotheses that are statistically tested, or, in a few cases, develops stories that lead to the formulation of simple mathematical models. It was Richardson's use of these two forms of analyses that brought me back to my original queries. The contrast clearly posed the question: What was the difference – if any – between the statistical test of a hypothesis and the creation of a mathematical model?

This question became increasingly pressing as statistics began to permeate the discipline. The field's focus on data collection and hypothesis testing necessarily required decision rules that could provide guidelines for rejecting hypotheses. Like many others, I joined the statistics bandwagon and began my education in this new set of tools. But my study and use of statistics increased my puzzlement. Statistics was mathematics. Did that mean that the application of statistics to data was mathematical modeling? Are chi square tests and correlations mathematical models? How did the equation for a correlation coefficient differ from the equations of the arms race model? Were the arms race equations a set of hypotheses that needed to be tested? Was there a difference between a mathematical model and a statistical analysis?

As I pondered these issues, I was asked to write a review of *Mathematics and Politics* by Alker (1965). Without explicitly noting the difference between a statistical analysis and mathematical modeling, Alker's survey of both made the contrast between the two explicit. The difference became obvious. The goal of statistics was to make coherent decisions: Did the data support the hypothesis? But the goal of mathematical modeling was to tell a story: How does an arms race begin and evolve? They were both mathematical enterprises, but their purpose was very different.

I returned to *Statistics of Deadly Quarrels* in an effort to better understand this difference. Many of Richardson's questions were simple hypotheses, and so he used his dataset together with statistical tests to confirm or reject. A few others, however, led him to formulate stories about a process he believed underlay the answer to the question. I found one such story to be of particular significance. It was a story about how nations might form alliances to fight a war. This story was especially valuable because Richardson used probability theory – combinations and permutations – a form of mathematics I understood. For the first time, I was able to follow Richardson's translation from a verbal story into its mathematical counterpart and witness the emergence of a testable conclusion.

While the story was too simple to be believable, its very simplicity made it possible to observe the mathematics in action. As Richardson put it, if his story were true – a phrase often used by Lave & March – then one would observe a specific distribution of the number of wars over the number of nations on either side: i.e. the number of wars in which one nation fought one nation, the number of wars in which one nation fought two nations, etc. Thus, the simple story about how nations formed alliances necessarily implied that types of wars (one nation against another, one against two, etc.) would result in a particular, i.e. predicted, distribution. The predicted distribution was a hypothesis and as such could be tested by comparing it to the distribution found in Richardson's dataset. For the first time, I saw the critical link between stories, mathematical modeling, and hypotheses. The translation of a story into a mathematical representation could lead to testable hypotheses.

The bits and pieces of answers to the questions from my graduate school days were emerging: the relevance of an initial question, the role of story-telling, the importance of mathematics, the difference between mathematical models and statistics. But gaps still remained, and the pieces still did not fit together. Where do the questions that initiate stories come from? How are stories developed? What information is needed to tell a story? How do you translate a story into mathematics? What is the difference between testing a hypothesis or telling a story, translating it into mathematics, and generating a hypothesis for testing? And finally, and most significantly, what happened to 'theory?'

In an attempt to fill the gaps, I returned to Richardson and Lave & March. Richardson had begun his arms race model with a story. Lave & March gave example after example of stories. But what constituted a 'story?' Merriam Webster proposes that a story is 'an account of incidents or events ... pertinent to a situation'. This definition suggested that a string of time-dependent sentences would qualify as a story if the sentences all referred to the same situation.

Trying out the definition, I constructed a 'story': 'I'm sitting in a coffee shop and a young lady enters, walks to the counter, and orders a cappuccino. She pays her bill, the waitress puts the money in the register, and makes the drink. The waitress hands the drink to the young lady who then goes to a vacant table, sits down, and enjoys her purchase.' According to the definition, I had created a story: a set of sequentially connected sentences concerning an incident. But despite the dictionary definition, the set of sentences didn't look like anything I would call a 'story'. I tried the exercise multiple times before becoming convinced that there was more to the 'story' concept than captured by the dictionary.

I headed back to Richardson's arms race and the Lave & March examples and discovered that I had overlooked a critical piece. Richardson's arms race story began with a question, the desire to understand the onset of war – he had a reason, a purpose, a question that he was attempting to understand. Likewise, Lave & March were curious about dorm friendship patterns or why all-women high schools provided a better education for college-bound women. In short, stories begin with a question. Thus, a story has a purpose: it is designed to answer a question, to explain why something happened. My sequentially connected sentences about the lady in the coffee shop was not triggered by a question, thus the set of sentences explained nothing. Perhaps if the young lady's boyfriend had been killed an hour earlier, the question might have been about the woman's potential complicity in the event. But in isolation, the coffee shop episode was of no interest, it explained nothing, it answered no question. It wasn't a story.

With this revelation in hand, I turned to storytelling by considering questions for which I wanted answers, events that I wanted to explain. But the task was daunting as I began by asking questions about wars, failed states, the reasons for revolutions. I quickly discovered that the questions were too big, or I didn't know enough to formulate a story to provide an answer. Then, taking a cue from Lave & March, I realized that if I looked around in my daily life I was frequently asking questions. They were tiny questions compared to a question about why wars occur, but they were questions that were nevertheless looking for answers. Moreover, they were questions for which I had enough information to construct a story. I decided to use these daily questions as a training ground to teach me how to recognize questions and formulate stories.

It quickly became obvious that lots of things were happening daily that I didn't understand, that didn't make sense. Why were there potholes in one part of town and not another? Why was grass growing alongside the road in a desert? Why did caterpillars congregate at certain intersections on country roads? Why was it so difficult to find a common time for three retired women to share lunch? As I spun stories about each question, I began to understand how questions arise. A question arises when an event occurs that contradicts what is expected. Potholes are the consequence of erratic temperatures during rough winters, but this should happen randomly throughout town. Since deserts receive little rain, how is it possible for grass to grow along a road? Caterpillars have limited cognitive abilities, so why are there congregations at 'intersections'? Retired folks no longer have work commitments, so why is it so difficult to find a mutual free time to meet for lunch? Richardson's story about an arms race was similar. His question was why war occurred when statesmen in opposing nations were attempting to prevent war through armament buildups?

I had begun to understand how questions arise and stories were developed. But I was still unclear about the use of mathematics. While I was convinced that translating a story into mathematics could produce a testable hypothesis, I was perplexed as to how to make the translation and use the mathematics to generate hypotheses. Then serendipity stepped in.

A colleague asked me to join forces on a methods textbook. The topics of the proposed text were to be typical, e.g. survey research, experimental design, etc. However, my colleague suggested that we approach the material from a very different perspective. He proposed that the text focus on questions about political processes and provide answers to those questions by constructing stories similar to those found in Lave & March. The stories would then be translated into the mathematics of basic logic – propositional calculus – and using the mathematics of propositional calculus, conclusions (hypotheses) could be generated. The resultant hypotheses would then be tested with a given methodology (e.g. survey research). To illustrate the process, my colleague had written a variety of political stories about congress (his area of expertise), translated them into propositional calculus, and using the mathematics of propositional calculus drawn some intriguing conclusions.

The methods text never materialized (too avant-garde for the publisher), but my colleague's examples became the critical final step towards answering my questions. Richardson's probability model provided the insight into how a mathematical translation occurs and how the mathematics of probability could generate a testable hypothesis. But the language of probability theory seemed too limited to be useful for stories about politics. Propositional calculus, however, appeared both accessible and useable. It was time for more study.

Perhaps the most important thing to come from my dive into the rudiments of propositional calculus, was the insight it gave me into the significance and meaning of theorems. Understanding the power of theorems led to my discovery of the home of the rabbits. I learned that everything is implicit in the original definitions, assumptions, and the basic rules of a form of mathematics. These are the building blocks needed to prove theorems. Theorems are just restatements of combinations and permutations of the definitions and assumptions given the accepted mathematical rules. The conclusion of a theorem is a restatement of the original assumptions. Theorems are then used to prove more theorems. Thus, it is always there – the rabbit is in the initial definitions and assumptions. A 'new' rabbit is the original one wearing different clothes. By translating a story into mathematics, it is possible to use theorems that point to 'new', i.e. implicit, information embedded in the assumptions that constitute the statements of a story. This is the power and value of mathematical languages. You begin by accepting as 'true' a basic set of premises and when you sign that contract you are given a panoply of consequences through theorems that show you all the other things that are then 'true'.

Propositional calculus is a very primitive form of mathematics compared to differential equations, but the underlying logic is the same. Theorems begin with definitions and assumptions that are accepted as 'true' and proceed to demonstrate that, given the rules of mathematics, another set of things are also true. Richardson's deductions about 'war' and 'peace' are implicit in the mathematics of differential equations used to capture the story of an arms race. Richardson's conclusions about 'war' and 'peace' come from the mathematics of differential equations. Thus, the link between a story translated into mathematics, and a generated hypothesis is transparent to anyone trained in that form of mathematics. This cannot be said about conclusions from purely verbal stories. This was the difference between Richardson's analyses and the verbal conclusions of Lave & March.

As I learned the specifics of propositional calculus and followed the examples of my colleague's political stories, the translation and hypothesis generation processes became clearer and the value of mathematics more obvious. Translation forces one to identify the key components of a story and the principle links between the components. This makes the outlines of a story obvious. The use of theorems to unambiguously draw conclusions links the theoretical world to the empirical. Moreover, drawing hypotheses from mathematical models can produce insights not seen otherwise.

To briefly see how mathematical modeling might work, let us consider Richardson's story about two statesmen in two neighboring states who are concerned about the possible intentions of the other. To translate this story into the mathematics of propositional calculus we need two sets of definitions. The first is the concept of an 'atom', defined as a simple statement that can be either True or False. In the arms race model, we can identify the following 'atoms:' Using symbols to represent the atoms we define:

X = the head of state X wishes to protect state X
A = the head of state X puts considerable resources into armaments
Y = the head of state Y wishes to protect state Y
B = the head of state Y puts considerable resources into armaments

It is easy to see that each of these statements could be assigned the value of T or F, e.g. the head of state does wish to protect state X (i.e. statement is T) or the head of state X does not wish to protect state X (i.e. the statement is F).

The second component of propositional calculus is the set of four 'operators' that link atoms to produce compound statements:

'and' ($\wedge$)
'or' ($\vee$)
'not' ($\sim$)
'implies' ($\rightarrow$)

Thus, the first part of the story might be translated into:

X $\rightarrow$ A

If the head of state X wishes to protect state X, then the head of state X puts considerable resources into armaments and

Y $\rightarrow$ B

If the head of state Y wishes to protect state Y, then the head of state Y puts considerable resources into armaments.

To continue the story, we define

C = state Y feels threatened
D = state X feels threatened

and then construct the compound statements:

A $\rightarrow$ C

If the head of state X puts considerable resources into armaments, then state Y feels threatened and

B $\rightarrow$ D

If the head of state Y puts considerable resources into armaments, then state X feels threatened.

To make the story simple let's define

E = state X declares war on state Y
F = state Y declares war on state X

Then C $\rightarrow$ F

If state Y feels threatened by state X, then state Y declares war on state X

D $\rightarrow$ E

If state X feels threatened by state Y, then state X declares war on state Y.
Finally, we define

W = states X and Y go to war.

and propose the compound proposition

$(F^\wedge E) \rightarrow W$

If state Y declares war on state X and state X declares war on state Y, then states X and Y go to war.

The theorems of propositional calculus tell us that if we begin with the atoms X and Y, together with the above story, we can conclude that the two states will go to war. Namely, given X and Y, i.e. two states with statesmen that wish to protect their state by putting resources into armaments, then W is a consequence, i.e. these two states will go to war.

Another intriguing conclusion that emerges from this story is the following: Using theorems from propositional calculus and skipping a few steps, the following can be concluded:

$$\sim W \rightarrow \sim (F \wedge E) \rightarrow (\sim F \vee \sim E) \rightarrow (\sim D \vee \sim C) \rightarrow (\sim B \vee \sim A)$$

This deduction says that if the story is true, then it should also be the case that when there are two (neighboring) states that have not gone to war it must be the case that at least one or both of those two states did not put considerable resources into armaments.Clearly, the above translation is overly simplistic. Like the translation of a text from one language to another, the process of translating a verbal story into a mathematical language is more an art form than a science. For any given verbal story there are many possible mathematical translations. We could for example have made the story more complicated by having only one state declare war and then the other retaliate. In this particular case, the conclusions are unlikely to be very different. However, it will be the case that different representations can lead to very different conclusions.

Propositional calculus was essentially the last step in my journey. I had the pieces needed to answer my graduate school questions. Stories are at the heart of theories. Theories, like stories, begin with a question and are designed to provide an answer to the query; theory/stories explain something. Casting a story in the language of mathematics – mathematical modeling – makes it possible to unambiguously produce conclusions, i.e. hypotheses, (deductions) that provide new insights and may be empirically verified. Empirical tests of deductions using statistical decision rules provide support for or against the original story. Theory and mathematical modeling are not equivalent; the latter provides a medium for evaluating the former. Mathematical modeling and statistical analyses are both mathematical enterprises, but they are used towards different ends. Mathematical modeling is an aid in the story-telling process while statistical analyses can provide the rules for empirical evaluation of the story. The story comes first and then its mathematical restatement provides the tool that produces new insights (hypotheses) to be evaluated. Data and statistics then follow to determine the empirical validity of the hypothesis and thus the grounds for determining the viability of the story.

Richardson had a profound effect on my life and career. *Arms and Insecurity* pushed me to learn mathematics and, through various twists and turns, eventually to an understanding of the power of a mathematically written story. *Statistics of Deadly Quarrels* provided both insights into data collection and measurement as well as a deeper understanding of the mechanics of mathematical modeling. Thus, the two volumes together gave me the basis for finally answering the questions of graduate school days.[5]

References

Alker, Hayward (1965) *Mathematics and Politics*. New York: Macmillan.

Azar, Edward E (2009) *Conflict and Peace Data Bank (COPDAB), 1948–1978*. Ann Arbor, MI: Inter-university Consortium for Political and Social Research, www.icpsr.umich.edu/icpsrweb/ICPSR/studies/07767.

Gillespie, John V; Dina A Zinnes, Gurcharan S Tahim, Philip A Schrodt & R Michael Rubison (1977) An optimal control model of arms races. *American Political Science Review* 71(1): 226–244.

Guetzkow, Harold et al. (1963) *Simulation in International Relations: Developments for Research and Teaching*. Englewood Cliffs, NJ: Prentice-Hall.

Lave, Charles A & James G March (1975) *An Introduction to Models in the Social Sciences*. New York: Harper & Row. [Reprinted 1993: Lanham, MD: University Press of America.]

McClelland, Charles A (1978) *World Event Interaction Survey*. Ann Arbor, MI. Inter-university Consortium for Political and Social Research, www.icpsr.umich.edu/icpsrweb/ICPSR/studies/5211.

Rapoport, Anatol (1957) Lewis F Richardson's mathematical theory of war. *Conflict Resolution* [now *Journal of Conflict Resolution*] 1(3): 249–299.

Richardson, Lewis F (1960a) *Arms and Insecurity: A Mathematical Study of the Causes and Origins of War*. Pittsburgh, PA: Boxwood.

Richardson, Lewis F (1960b) *Statistics of Deadly Quarrels*. Pittsburgh, PA: Boxwood.

Rummel, Rudolph J (1972) *The Dimensions of Nations*. Beverly Hill, CA: Sage.

Singer, J David (1969) The incomplete theorist: Insight without evidence. In: Klaus E Knorr & James N Rosenau (eds) *Contending Approaches to International Politics*. Princeton, NJ: Princeton University Press, 110–128.

Singer, J David & Melvin Small (1972) *The Wages of War, 1816–1965: A Statistical Handbook*. New York: Wiley.

Zinnes, Dina A (1980) Three puzzles in search of a researcher. Presidential address. *International Studies Quarterly* 24(3): 315–342.

---

[5]For examples of how I drew on Richardson in my own work, see Gillespie et al. (1977) and Zinnes (1980).

Dina A. Zinnes, b. 1935, Ph.D. in Political Science (Stanford University, 1964); Professor of Political Science, Indiana University (1967–80), Merriam Professor of Political Science, University of Illinois (1980–2005), now emeritus. Founded and directed the Merriam Laboratory for Analytic Political Research (1986–2005). President of the International Studies Association (1980–81); Editor, *American Political Science Review* (1981–85), President, Peace Science Society (1989), zinnes@illinois.edu

Chapter 3
The Influence of the Richardson Arms Race Model

Ron P. Smith

Abstract This chapter reviews the Richardson arms race model: a pair of differential equations which capture an action reaction process. Whereas many of Richardson's equations were quite specific about what they referred to, the arms race model was not. This lack of specificity was both a strength and a weakness. Its strength was that with different interpretations it could be applied as an organising structure in a wide variety of contexts. Its weakness was that the model could not be estimated or tested without some auxiliary interpretation. The chapter considers the impact of these issues in interpretation and empirical application on the influence of the Richardson arms race model.

3.1 Introduction

There are many definitions of arms races, but for the purpose of this chapter they can be thought of as enduring rivalries between pairs of hostile powers which prompt competitive acquisition of military capability. Two approaches to modelling arms races have been particularly influential. One is as a two-person game, in particular the Prisoner's dilemma, where the choices are to arm or not to arm, and the dominant strategy, for both to arm, is not Pareto optimal. The other, which is the focus of this chapter, is the Richardson model of the arms race as an action-reaction process, represented by a pair of differential equations.

Just as the two supply and demand equations have structured thought about the dynamics of markets for most economists, the two Richardson equations have structured thought about the dynamics of arms races for most subsequent analysts. Not only did he develop the model, he attempted to test it using data on military expenditure prior to World War I. One of the strengths of the model is that it has prompted a range of questions, many of which Richardson himself posed. This chapter reviews the influence of the Richardson arms race model on the subsequent literature through these questions, which include: What are the characteristics of the

This is a revised version of a paper presented at the Richardson Panel at ISA 2018, Saturday 7 April 2018. I am grateful to participants of the panel for comments.

N. P. Gleditsch (ed.), *Lewis Fry Richardson: His Intellectual Legacy and Influence in the Social Sciences*, Pioneers in Arts, Humanities, Science, Engineering, Practice 27, https://doi.org/10.1007/978-3-030-31589-4_3

solution to this model? What are the variables and actors? What is the time dimension? How do arms races relate to wars? How should the parameters be interpreted? How can the model allow for strategic factors and budget constraints? How should the model be related to the data? How do you stop arms races?

Given the variety of ways that his arms race model has been used, I was tempted to call this piece 'variations on a theme by Richardson', but specifying the theme precisely proved problematic. The voluminous arms race literature that arose from his work has many themes. In trying to identify the themes, I found the papers in the collection edited by Gleditsch & Njølstad (1990), hereafter G&N, very useful. G&N provides an overview that lies roughly half way between the publication of Richardson (1960a) which brought his arms race model to a wider audience and the present day. With the end of the Cold War, interest in arms races declined somewhat and many of the themes that are in that book remain central. It is difficult to say anything new in this area, and Wiberg (1990) makes many of the same points as I make below. There is a more technical discussion of many of the issues mentioned here in Dunne & Smith (2007).

As most readers of this chapter will probably know, Lewis Fry Richardson (1881–1953) was a Quaker physicist, a Fellow of the Royal Society, who made major contributions in the mathematics of meteorology, turbulence and psychology as well as his work on quantifying conflict. The significance of the work on conflict was only widely recognised posthumously with the publication of Richardson (1960a), *Arms and Insecurity*, which introduced the arms race model, and Richardson (1960b) *Statistics of Deadly Quarrels,* which looked at the distribution of conflict deaths.

Richardson was a very careful scientist. When he was investigating the hypothesis that the probability of war between two countries was a function of the length of their common border, he double-checked the data and noticed that adjoining countries gave different lengths for their common border: the smaller country tending to think the border longer than the larger. This was because the measured length was a function of the size of the ruler or scale of the map used; small countries tended to use smaller rulers and larger scale maps. Richardson's subsequent studies on this phenomenon introduced the idea of non-integer dimensions and prompted Mandelbrot's work on fractals. A common border may increase the probability of war but, as Richardson recognised, it also tends to increase the amount of trade, which may have a pacifying effect.

Richardson approached the analysis as a physicist. He often used differential equations to characterise the dynamics, and tried to match the models to data, often using probabilistic techniques. His work provides excellent teaching material in applied mathematics. Students find the arms race model a nice motivation for a neat system of differential equations which has a range of interesting solutions. Korner (1996) makes pedagogical use of a number of examples of Richardson's work to motivate the applications of mathematics, as well as discussing Richardson's life and influence. The teaching aspect is also noted in a recent paper, Beckmann, Gattke, Lechner & Reimer (2016: 22–23), say about the Richardson equations: 'our objective was to see whether this old staple can be brought back from the world of teaching (where it serves as an example for solving systems of differential equations) into modern research on conflict dynamics.'

While serving in the Friends Ambulance Unit during World War I, Richardson began to try to describe the causes of war in systems of equations, which he published as Richardson (1919). Maiolo (2016: 1) says the term arms race originated in the 19th century and was commonly presented as one of the causes of World War I. He quotes Lord Grey, who had been Foreign Secretary when Britain went to war, as writing after the war that 'Great armaments lead inevitably to war. If there are armaments on one side, there must be armaments on the other sides' (Maiolo, 2016: 2). Richardson also cited Grey and thus his equations captured a common perception of the cause of that war. However, as Maiolo (2016: 4) also notes, the sporting metaphor can be misleading: in athletics races have clear start and finish lines, arms races do not.

Richardson was not alone in trying to develop mathematical models of conflict. About the same time, Lanchester (1916), based on articles published in *Engineering* in 1914, developed models of the evolution of different types of battle. The models examined the role of the quantity and quality of forces deployed. Lanchester also used a pair of differential equations though to different ends. One might distinguish a Lanchester tradition, in operational research, of mathematical modelling to win wars, from a Richardson tradition, in peace research, of mathematical modelling to stop wars. MacKay (2020, in this volume) discusses a combination of Richardson's arms race equations with Lanchester's attritional dynamics.

3.2   The Equations

The Richardson model describes the path over time, t, of the level of arms, x and y, of two countries, A and B.

$$dx/dt = ky - \alpha x + g$$
$$dy/dt = lx - \beta y + h$$

The rate of change of the arms of each country is the sum of a positive reaction to the arms of the other country, a negative reaction to the level of its own arms through a 'fatigue' factor and a constant component through a 'grievance' factor. Setting it up in this way prompts a set of questions which are internal to the mathematical structure of solving linear differential equations. Does an equilibrium exist? Is it unique? Is it stable? Are there boundary conditions, e.g. x, y > 0?

In equilibrium dx/dt = dy/dt = 0, so the equilibrium reaction functions are two straight lines

$$0 = ky - \alpha x + g$$
$$0 = lx - \beta y + h$$

If $\alpha\beta$ = kl the lines are parallel, otherwise they intersect once at an equilibrium, which may involve negative values. Because of linearity, if the equilibrium exists, it is unique, and one can then consider how stability varies as a function of the parameters. Here arms race stability refers to the nature of the solution to these equations. An unstable solution would diverge from an equilibrium, for instance exhibiting exponential growth by both countries. Arms race stability is not the same thing as strategic stability, which can itself have many meanings. Richardson related them by suggesting that exponential growth could lead to war, though in principle it could lead to bankruptcy. Diehl (2020, in this volume) discusses the links between arms races and war.

Again, within the internal mathematical structure it is natural to ask if the model generalises. What happens if there are three or more actors? What happens if one relaxes the assumption of linearity? There is a large literature on both these questions. Broadly, as in the three-body problem in physics, the neat simplicity of the conclusions is lost when the model is generalised and multiple equilibria may exist. For instance, among three countries, the equations for each pair of nations may be stable, but the triplet is unstable.

The model has an immediate common-sense plausibility as a description of an interaction between hostile neighbours. This is what makes it such a nice teaching tool. There are also historical examples of such reaction functions, for instance the British policy before World War I of having a fleet as large as the next two largest navies combined. But the model has no unambiguous interpretation. In the physical sciences, when Richardson used equations, for instance in fluid dynamics, he knew exactly what the variables were, what measures they corresponded to, and the time dimension of the dynamic processes involved. Little interpretation was needed. But in the social sciences the interpretation of mathematics is rarely unambiguous.

3.3   Interpreting the Equations

The arms race equations prompted a number of questions about the interpretation. There were questions about how to interpret the measures of arming, x and y. In a symmetric arms race, they were the same variable, such as military expenditures or number of warships. In an asymmetric arms race, they could be different types of variable; historically there was an arms race between castle design and siege train technology. They might be quantitative, number of warheads, or qualitative, accuracy of the missiles. They might be given a more psychological interpretation as hostility or friendliness. There were many possibilities.

There was also a question about how to interpret the nature of the actors, A and B, and the motives for their actions. They might be countries, alliances, decision makers or non-state actors like terrorists. Their actions might be the result of rational calculations or bureaucratic rules of thumb and there were many possible sources of their hostility. Some, like Intriligator (1975), felt the need to motivate the equations with an explicit objective function for the actors. There were also questions about the time period, months or centuries, over which the interactions were taking place and the extent to which the parameters could be regarded as stable. Finally, from the policy perspective there was the crucial question: how might you stop the process?

This lack of specificity was both a strength and a weakness. Its strength was that the model could be applied in a wide variety of domains, by giving the variables x and y and the actors A and B different interpretations. As it was imported into a particular domain, other questions would arise. For instance, an economic interpretation would immediately prompt questions about the nature of the budget constraint. Economists tended to allow for the budget constraint by adding income as an extra variable, but there were many other ways, for instance Wiberg (1990: 366–367) assumes a fixed amount of resources available.

The weakness of the lack of specificity was that there was little clarity either about the precise predictions of the model or about the evidence that would falsify it. As a specific example, the parameters α and β could be interpreted as representing: (a) a measure of fatigue, as Richardson did: increased spending exhausts a country depressing the growth of arms; (b) the speed of adjustment towards a desired level in a stock adjustment model or (c) a measure of bureaucratic inertia; or perhaps some combination of the three. The form of the equation would be identical, but the story one told about the parameters would be different in each case. This was important, since in practice these parameters were estimated statistically and needed interpretation. If one does not know where the parameters came from or why they might differ, between the countries or over time, it is difficult to judge whether the statistical estimates are sensible.

Just as the term arms race is a metaphor, any model is a metaphor (the equations are interpreted as being like the world in some respects) and there is an issue as to how literally to take these equations. Some, like Beckmann et al. (2016) in a critique of the Richardson equations, treat them very literally. If they do not hold exactly, then the Richardson model is wrong or it is a different model. Others treat the model as being more loosely defined and are happy to label any set of equations involving action-reaction processes as a Richardson type model. Intriligator (1975) and Dunne & Smith (2007) take this approach. Economists, treating them like supply and demand curves, organising principles rather than exact specifications, seem inclined to take them less literally.

Of course, some do not accept the action-reaction description itself. Senghaas (1990: 15) rejects the explanation of the arms race as an other-directed reciprocal escalation spiral: 'As much research on the biography of weapons systems has shown, the action-reaction scheme is at least highly dubious, if not completely false.' Instead he sees it as inner-directed by the self-centred imperatives of national armaments policy. Gleditsch (1990: 8–9) lists a very large number of explanations for arms acquisitions, organised under four levels: (1) internal factors, such as particular interest groups; (2) actor characteristics such as being an alliance leader or authoritarian rule; (3) relational characteristics, such as action reaction or relations to allies and (4) system characteristics, such as upswings in long economic waves and technological imperatives. While the focus in this chapter is on action-reaction explanations of arms races, much of the work on other explanations was prompted by the desire to criticise the Richardson action-reaction explanation.

On whose behaviour they described, Richardson (1960a: 12) was enigmatic about the interpretation of the equations: 'the equations are merely a description of what people would do if they did not stop to think'. Intriligator (1975) derives Richardson type equations from optimising strategies in a nuclear war. Brito & Intriligator (1999) argue that new military technologies which imply increasing returns should mean the end of the Richardson paradigm with its implicit assumption of constant or declining returns to scale. The behaviour of participants and the research questions in increasing returns to scale systems are very different. For instance, multiple equilibria are possible, and arms control may have the potential to move the system from a high to low equilibrium. Increasing returns to scale increases the dominance of the dominant actor in its chosen technology, providing incentives for the non-dominant actors to choose alternative technologies such as terrorist attacks.

Relating the equations to data

Richardson evaluated the model through an examination of the growth in military expenditures, 1908–14, of the two belligerent alliances, the Entente and Central powers, prior to World War I. He took the observed exponential growth as an indication of support for his models. He interpreted x and y as measuring military expenditures, A and B as coalitions, and the relevant time period as 7 years. However, he noted that other conflicts were not preceded by arms races.

As has been widely noted, e.g. by Gleditsch (1990: 9–10), there is an identification problem: quite different models can give observationally equivalent predictions. While one solution of the arms race model is exponential growth, exponential growth may equally well result from purely internal processes within each country, such as a military industrial complex, with no action-reaction component. Exponential growth may also result from both countries responding to a third country. Expectations further complicate the matter as discussed in Dunne & Smith (2007).

The empirical literature separated into a number of separate tracks. One track looked at whether arms races, suitably defined, preceded conflicts, again suitable defined. Diehl (2020) reviews this track. Another track looked at estimating the Richardson equations directly to see whether they showed action reaction features: significant coefficients for the arms of the other countries. This was usually done from time series though there are also some cross section and panel papers looking at arms race interactions.

To estimate the Richardson equations directly from time series data, they required various modifications. The equations had to be converted from continuous into discrete time, with corresponding judgements about the time-scale involved, how many lags were required and the interval over which one might expect the parameters to be stable. Typically, the lagged dependent variable, arms in the previous period, is a very strong predictor of the current value.

Specific measures had to be chosen for x and y. The logarithms of military expenditures and the shares of military expenditure in GDP were popular choices, but there were many other possibilities, including physical measures like number of warheads. Even when using military expenditures, the estimates could be quite sensitive to other measurement issues, such as the choice of exchange rate used to make them comparable. Of course, expenditures are an input rather than an output, capability, measure. Countries may differ in their efficiency, the amount of military expenditure required to achieve a particular level of capability.

The equations are deterministic and had to be supplemented by stochastic specification. Typically, 'well-behaved' error terms were added to the equations, but again there were many other possibilities, depending, for instance, on how one treated the endogeneity that resulted from the variables being jointly determined, serial correlation and heteroskedasticity. Supplemental variables might be added to control for other factors, e.g. GDP to allow for the budget constraint.

Given all these decisions, it could be difficult to judge what light these estimates threw on the Richardson equations. Firstly, as noted above, since Richardson provided little in the way of interpretation of the coefficients, it was not always clear whether the statistical estimates were consistent with his model or not. Secondly, there is the Duhem-Quine problem: any test involves a joint hypothesis. What is being tested is both the substantive hypothesis, the validity of the Richardson model in this case, and a set of auxiliary hypotheses, such as those about choice of measure, dynamics and functional form. One never knows whether it is the substantive or the auxiliary hypotheses that has led to rejection. McKenzie (1990) discusses the Duhem-Quine problem in the context of the sociology of nuclear weapons technologies. The converse of this problem is that since the Richardson model is not very specific, this allows great freedom for specification search over such things as measures for x and y; functional forms; dynamics; estimation methods; sample period and control variables included. This search can continue until one finds a specification that confirms one's prior beliefs.

Despite these qualifications most surveys of this literature including Dunne & Smith (2007) conclude that there is limited time-series evidence for stable equations, of the Richardson type, describing the interaction of quantitative measures of military expenditure or capability. That article discusses the case of India and Pakistan, where there had been more evidence of a stable Richardson type action-reaction process between constant dollar military expenditure, 1962–97, but it seemed to have broken down after 1997, about the time both powers went nuclear. Empirical estimates of Richardson type equations are sensitive to choice of measure of military expenditure and to many aspects of specification such as other covariates included and functional form used.

3.4   Other Arms Races

The arms race metaphor has spread beyond military interactions and a comparison with its use in another area is revealing. We changed the title of Dunne & Smith (2007) from the one we had been given 'The econometrics of arms races' to 'The econometrics of military arms races', because on putting the term 'arms races' into Google Scholar the top paper was Dawkins & Kreps (1979) 'Arms races between and within species', followed by many highly cited biological papers.

The comparison between Dawkins & Kreps and Richardson is interesting both for the similarity in the process and difference in approach: they are much more specific, much less metaphorical than Richardson. They do not cite Richardson but have a very similar process in mind: 'An adaptation in one lineage (e.g. predators) may change the selection pressure on another lineage (e.g. prey), giving rise to a counter-adaptation. If this occurs reciprocally, an unstable runaway escalation or 'arms race' may result' (Dawkins & Kreps, 1979: 489).

They begin using a military analogy and clarifying the time scales considered. 'Foxes and rabbits race against each other in two senses. When an individual fox chases an individual rabbit, the race occurs on the time scale of behaviour. It is an individual race, like that between a particular submarine and the ship it is trying to sink. But there is another kind of race, on a different time scale. Submarine designers learn from earlier failures. As technology progresses, later submarines are better equipped to detect and sink ships and later-designed ships are better equipped to resist. This is an 'arms race' and it occurs over a historical time scale. Similarly, over the evolutionary time scale the fox lineage may evolve improved adaptations for catching rabbits, and the rabbit lineage improved adaptations for escaping. Biologists often use the phrase 'arms race' to describe this evolutionary escalation of ever more refined counter-adaptations (Dawkins & Kreps, 1979: 489–490). They cite use of the term arms race in a biological context in a 1940 biology paper, though as noted above the term arms race goes back to the 19th century.

They are also specific about who is involved. 'In all of this discussion it is important to realize who are the parties that are 'racing' against one another. They are not individuals but lineages' (Dawkins & Kreps, 1979: 492). They distinguish between symmetric and asymmetric arms races, arguing that asymmetric arms races are more likely between species and symmetric ones within species, e.g. male-male competition for females. They propose the 'life-dinner principle': when a fox chases a rabbit, the fox is running for its dinner, the rabbit is running for its life. Thus, the incentives and the evolutionary selection pressures on the rabbit are greater. This principle has obvious military analogies in cases such as Vietnam where the weak defeat the strong, because the weak have more at stake. They do not have any equations in the paper, but much of the work they cite, such as by William D Hamilton and John Maynard Smith, is mathematical, often involving game theory, particularly evolutionary stable games.

Dawkins & Kreps are quite specific about the time scales, parties and mechanisms involved in their biological arms races. This is like Richardson's treatment of physical processes and statistics of deadly quarrels, but unlike his more metaphorical treatment of the mathematics of military arms races.

3.5 Conclusion

Military arms races are perceived as common and usually regarded as a bad thing. Wiberg (1990: 353) suggests that they are matters of concern because of the risk of war, the waste of resources, the threat to other states, and the danger that they can breed militarism. The two main tools we have for understanding the process have been game theory, particularly the Prisoner's dilemma, and the Richardson model. The Richardson model has motivated much more empirical work than game theory, where papers tend to just use illustrative historical examples to motivate the mathematics, rather than to attempt to test the theory.

The strength and the weakness of the Richardson arms race model was that it was not very specific. This was a strength in that by giving the variables and actors different interpretations, it could be applied in a wide variety of contexts and prompt a range of interesting questions. It was a weakness in that it made it difficult to evaluate the theory. Richardson's models for the distribution of conflict statistics, power laws for size and Poisson distributions for frequency, were more like physical results and have been widely replicated. It may be that arms races, representing historically specific human decisions, are not subject to systemic regularities, so being prompted to ask the right questions is helpful in itself.

As Gates, Gleditsch & Shortland (2016: 345) put it 'Richardson's formal dynamic model of arms races may not be very useful as a description of the data or as an explanation of conflict – indeed, no decision to use force per se appears in the model. Still it is clear that it has helped move the field ahead and stimulate new research and interest in formal models of conflict.'

References

Beckmann, Klaus; Susan Gattke, Anja Lechner & Lennart Reimer (2016) A critique of the Richardson Equations. *Economics Working Paper,* Helmut Schmidt University (162).

Brito, Dagobert L & Michael D Intriligator (1999) Increasing returns to scale and the arms race: The end of the Richardson paradigm? *Defence and Peace Economics* 10(1): 39–54.

Dawkins, Richard & John R Kreps (1979) Arms races between and within species. *Proceedings of the Royal Society of London, Series B, Biological Sciences* 205(1161): 489–511.

Diehl, Paul F (2020) What Richardson got right (and wrong) about arms races and war. Ch. 4 in this volume.

Dunne, J Paul & Ron P Smith (2007) The econometrics of military arms races. Ch. 28 in: Todd Sandler & Keith Hartley (eds) *Handbook of Defence Economics,* 2. Amsterdam: Elsevier, 914–941.

Gates, Scott; Kristian Skrede Gleditsch & Anja Shortland (2016) Winner of the 2016 Lewis Fry
    Richardson Award: Paul Collier. *Peace Economics, Peace Science and Public Policy* 22(4):
    338–346.
Gleditsch, Nils Petter (1990) Research on arms races. Ch. 1 in: Nils Petter Gleditsch & Olav
    Njølstad (eds) *Arms Races*. London: Sage, 1–14.
Gleditsch, Nils Petter & Olav Njølstad (eds) (1990) *Arms Races: Technological and Political
    Dynamics*. London: Sage.
Intriligator, Michael D (1975) Strategic considerations in the Richardson model of arms races.
    Journal of Political Economy 83(2): 339–353.
Korner, Thomas W (1996) *The Pleasures of Counting*. Cambridge: Cambridge University Press.
Lanchester, Frederick W (1916) *Aircraft in Warfare: The Dawn of the Fourth Arm*. London:
    Constable.
MacKay, Niall (2020) When Lanchester met Richardson: The interaction of warfare with
    psychology. Ch. 9 in this volume.
Maiolo, Joseph (2016) Introduction. Ch. 1 in: Thomas Mahnken, Joseph Maiolo & David
    Stevenson (eds) *Arms Races in International Politics: From the Nineteenth to the Twenty-first
    Century*. Oxford: Oxford University Press, 1–10.
McKenzie, Donald (1990) Towards an historical sociology of nuclear weapons. Ch. 8 in: Nils
    Petter Gleditsch & Olav Njølstad (eds) *Arms Races*. London: Sage, 121–139.
Richardson, Lewis F (1919) *The Mathematical Psychology of War*. Oxford: Hunt.
Richardson, Lewis F (1960a) *Arms and Insecurity: A Mathematical Study of the Causes and
    Origins of War*. Pittsburgh, PA: Boxwood.
Richardson, Lewis F (1960b) *Statistics of Deadly Quarrels*. Pittsburgh, PA: Boxwood.
Senghaas, Dieter (1990) Arms race dynamics and arms control. Ch. 18 in: Nils Petter Gleditsch &
    Olav Njølstad (eds) *Arms Races*. London: Sage, 346–351.
Smith, Ron P (1987) Arms races. In: John Eatwell, Murray Milgate & Peter Newman (eds) *The
    New Palgrave Dictionary of Economics*, vol. I, A to D. London: Macmillan, 113–114.
Wiberg, Håkan (1990) Arms races, formal models and quantitative tests. Ch. 2 in: Nils Petter
    Gleditsch & Olav Njølstad (eds) *Arms Races*. London: Sage, 31–57.

Ron P. Smith, b. 1946, Ph.D. in Economics (University of Cambridge, 1975); Professor of
Applied Economics, Birkbeck University of London; most recent book *Military Economics: The
Interaction of Power and Money* (Palgrave, 2009), r.smith@bbk.ac.uk.

Chapter 4
What Richardson Got Right (and Wrong) About Arms Races and War

Paul F. Diehl

Abstract This chapter considers Richardson's classic arms race model and his book *Arms and Insecurity* in relation to the association between arms races and war. The analysis begins with a short review of the academic debates and empirical research about arms races and war. This is a prelude to an examination of the dynamics of arms races as they are specified by Richardson in his book and in his arms race model. Following is an evaluation of the specific claims about arms races and war derived from the model as well as Richardson's own empirical analyses about the connection between the two phenomena. Although in light of subsequent research, a number of Richardson's arguments seem misguided or inaccurate, other insights were validated by later studies or remain understudied but worthy of future research.

Lewis Fry Richardson was one of the pioneers of the quantitative study of war (along with Quincy Wright and Pitirim Sorokin), and his work has stood the test of time and continues to influence scholars. Richardson is most famous for his equations that model the dynamics of arms races. Indeed, these dominate much of his book *Arms and Insecurity* (Richardson, 1960). Nevertheless, it is easy to forget that the subtitle of the work is *A Mathematical Study of the Causes and Origins of War*. Thus, there was a clear belief for Richardson that arms races, although not all of them, were associated with the outbreak of war. Periodically throughout the book, he refers to arms races and war, and includes a series of case studies of arms races in the 19th and first half of the 20th century.

How much of Richardson's assessment of arms races and war have turned out to be correct (or not)? Since the publication of his work, there has been an extensive set of research on the connection of these two phenomena, and we now have an empirical basis for reflecting on Richardson's insights at the dawn of systematic, empirical work on war. The focus of this chapter is to present Richardson's models and claims as they related to arms races and war, and then evaluate them in light of subsequent research.

© The Author(s) 2020
N. P. Gleditsch (ed.), *Lewis Fry Richardson: His Intellectual Legacy and Influence in the Social Sciences*, Pioneers in Arts, Humanities, Science, Engineering, Practice 27, https://doi.org/10.1007/978-3-030-31589-4_4

4.1   The Arms Race–War Debate

A debate over the dangers (or lack thereof) of arms races for the onset of war have
been among the most enduring in international relations scholarship. Although
Huntington (1958) first raised the issue, controversy did not ensue until the publication of Wallace's (1979) article more than two decades later. This article
received extensive attention in the scholarly and policymaking communities not the
least for its relevance to the nuclear disarmament campaign at the time and its
provocative findings. Wallace essentially argued that arms races were extremely
dangerous and had strong likelihood of leading to war between the participants. The
study was severely flawed and not replicable, and consequently it attracted a fair
amount of criticism for its methodology and case selection (e.g., Houweling &
Siccama, 1981; Diehl, 1983).

The original Wallace article had the benefit of raising the issue of the arms
race-war connection, but also the unfortunate side effect of focusing the debate on
research design concerns rather than theoretical explanations. The arms race-war
debate has settled down somewhat with a consensus emerging that the competitions
are dangerous in some contexts and under some circumstances (see Sample, 2012
for an overview), although whether they are causal or merely symptomatic is still
contentious (Rider, Findley & Diehl, 2011). The work of Richardson (1960) was
often perfunctorily cited in these debates, but rarely if ever were his specific ideas
about the arms races and war discussed or incorporated in the analyses. The discussion below is designed to return to Richardson's original formulation to correct
that deficiency and determine whether the literature has missed anything that would
be of value to future research.

4.2   Richardson on Arms Races and War

For a work that is purportedly about the origins of war, *Arms and Insecurity* has
remarkably little to say directly about the conditions for the outbreak of war,
especially as they relate to arms races. Indeed, the book is largely dedicated to
modelling the dynamics of arms races, which is also the focus of Smith (2020, in
this volume). The famous differential equations for arms races and their variations
take up most of the work. Not surprisingly then, most of the research that has been
influenced by (and cites) Richardson's work has been dedicated to arms race
models, even as those models have some notable limitations (see Zinnes, 1980). To
the extent that empirical analyses are found in his work, they are focused more on
fitting the models to data on military expenditures than on linking arms races to war
in a causal process. Case studies of World Wars I and II are sprinkled throughout
the book, and these are no doubt chosen because they were contemporary examples
for Richardson, but their magnitude also commanded attention.

Trawling through the work, there are numerous insights about arms races and war. Some of these come from the differential equations that reviewed by Smith (2020). Nevertheless, more are found in the narrative and case studies that make up the rest of the book. In sorting these out, I have roughly divided these into three categories below. The first set covers the dynamics of arms races as they relate to the study of arms races and war. The second deal with specific claims about the connection between the two phenomena. Finally, there are empirical analyses, primary and secondary, reported by Richardson.

4.2.1 The Dynamics of Arms Races and Their Implications for the Study of War

Richardson was prescient in a number of ways on how scholars should look at arms races and the conditions for war more broadly. Several of his ideas portend later, and more precise, formulations in the conflict studies literature. A number of the insights specifically about arms races remain salient and somewhat unexplored today. Richardson was also very concerned about data quality and measurement, specifically the reliability and comparability of the data were as well as the match between concepts and indicators. Although this chapter does not cover these issues, the reader is encouraged to examine and emulate the good practices followed by Richardson.

A repeated theme in the narrative is a concern for considering the power distribution of the states involved and whether each had a status quo versus revisionist orientation (Richardson, 1960: 26, see also 13, 35). This is now evident in light of empirical research on war over the last 70 years, but at the time (contemporary to Morgenthau and realism), this was an important formulation. With respect to arms races, Richardson (1960: 21) claims that the defense coefficient is proportional to the size of the state, indicating that extant capabilities and resources matter. There is also the related concern that defense burdens are important, both for the pace of the arms race and for the timing of the outbreak of war (more on this below).

Despite what might now seem as obvious considerations, much of the arms race-war empirical work ignores these considerations, even as these are centerpieces of general theories of war (for a general review of the literature, see Sample, 2012). For example, there is a tendency to look at the rates of increase in military spending for each side in an arms race without consideration to how the power distribution affects those rates, or vice-versa. There is also little consideration given to how defense burdens constrain (part of the fatigue factor in Richardson, 1960: 13), or facilitate arms increases and their relation to the outbreak of war (an exception is Diehl, 1985). Such failures have led much of the arms race-war literature to be delinked from the rest of the empirical conflict literature on power transitions and other theoretical models.

In a related fashion, Richardson makes reference to what we call 'rivalries' and how these influence arms races. The rivalry element suggests that we look at arms races in that context, a point raised in some recent debates about arms races and war (Rider, Findley & Diehl, 2011; Sample, 2012). Doing so brings in the issues of relative power distribution and status quo-revisionist considerations noted above. Rivalry can also be one element of what Richardson calls 'grievances', one of the drivers of arms increases in his model; grievances is a polyglot category encompassing all factors independent of the opponent's expenditures in driving one's own arms increases (Richardson, 1960: 16). This comports with rivalry research that finds that the 'pull of the past' and prior rivalry interactions influence current conflict behavior (Goertz, Jones & Diehl, 2005).

Richardson also captures the elements of what we now refer to as the 'security dilemma' (Jervis, 1978a, b) when he states that '… governments which develop their armaments cannot expect their professions of peace intentions to be believed …' (Richardson, 1960: 65). Mistrust and hostility coming from threat perception is the theoretical basis for scholarly arguments that posit that arms races increase the likelihood of war (Sample, 2012). Arms races increase the perception of threat, and indeed it is the reactivity of the Richardson model that captures this, and this undermines the ability and willingness of states to resolve their disagreements peacefully.

Another aspect of arms race dynamics put forward by Richardson that could influence its connection to war is the existence of cooperation between arming states. This could mitigate the impact of arms races (Richardson, 1960: 29ff, Chs VIII, XX). Trade is treated as the indicator of cooperation for Richardson's purposes, but his point is broader than about the effect of this particular variable (Richardson, 1960: 32). Although there is a large and vibrant literature on the impact of trade and interdependence on war (Schneider, 2010), countervailing factors are not usually a part of the arms race-war literature.

Richardson offered several propositions that subsequently have or should have influenced the arms race-war literature, but there are other aspects of his dynamics of arms races that are suspect for understanding their connection to war. One is how one treats aligned states in an arms race context. Richardson assumes that all allies are part of the arms race dynamics on the given side; in practice, this means aggregating their military expenditures (Richardson, 1960: 31f). The justification for this is atheoretical and especially weak: '… an important advantage to be gained by lumping together the statistics from as many nations as can be reasonably treated; for this agglomeration smooths the statistics and moves them further away from the domain of free will toward the domain of exact science' (Richardson, 1960: 80). He does back away from this later in the book when he raises the concern that one must pay attention to whom is targeted by the increases (Richardson, 1960: 159ff). The arms race-war literature has struggled with the problem of disaggregating war initiations (e.g., World War I) into dyads (see Wallace, 1979; Diehl, 1983), but has not generally aggregated alliance spending when calculating arms race scores. To do so masks individual state decisions for spending and war, implying that allies act as one entity and are perceived as such by enemies; nevertheless, we know that alliances are not all reliable (Leeds, 2003). Richardson's claim that after alliances the defense coefficients of allies vis-à-vis each other go to zero (Richardson, 1960: 170) seems more of an unwarranted assumption than a reflection of cases in which allies are also rivals with each other (e.g., Greece-Turkey) or fight wars against one another (Ray, 1990).

4.2.2 More Specific Claims About Arms Races and War

At various junctures in the book, Richardson makes references to arms races and war, albeit sometimes cryptically and others in ways that make it difficult to discern when arms races are dangerous and when they are more benign. Indicative of this is late in the book when he refers to have presented a 'theory of the stability or instability of peace' (Richardson, 1960: 147). One might guess that he is referring to negative peace, or the absence of war, but this is perhaps the first time Richardson makes explicit reference to the conditions for peace.

The most direct discussion of arms races and war comes in Ch V in the subsection labelled 'Bankruptcy or War?' (Richardson, 1960: 61f). Starting with his famous equations, Richardson notes that x and y (the increases in defense expenditures) would go to positive infinity, as the action-reaction process was deterministic. He notes that a critic (Professor Piaggio) argued that 'An infinite cost of armaments is interpreted as denoting war, though it might have seemed more natural to have taken it as bankruptcy' (quoted in Richardson, 1960: 61). In a strict interpretation, neither the bankruptcy nor war is a plausible outcome from the arms race process as specified. As Richardson correctly notes, there is no known case of a state going bankrupt from an arms race, a statement as true when written as today. More significant is the imprecision as to what point war becomes the outcome of the arms race. Is it only when one or both states have exhausted all their resources in pursuit of building armaments? This seems absurd and equally without empirical referents.

Richardson's other responses to his critic offer somewhat better, albeit still flawed, clues as to when arms races lead to war (Richardson, 1960: 61–62). Least compelling is the idea that '… diplomatic relations become, during the tacit mutual threats of an arms race, such a strain that the outbreak of war is felt as a relief' (Richardson, 1960: 61). This suggests a threshold for war short of resource exhaustion, but it is not clear when that might be or how it might vary across states and their leadership. Nevertheless, the crippling effects of resources devoted to military spending is the basis for Huntington's (1958) assertion that quantitative arms races are destabilizing and prone to war, especially as they lengthen; this is consistent, although no more precise in terms of timing, with Richardson's formulations.

More useful is the implication that arms races have a strong psychological impact on decision makers, and this maps well with Richardson's emphasis on threats and the aforementioned security dilemma argument. Indeed, Richardson

argues that the x and y variables in the structural equations of arms races conceptually represent threat and that military expenditures are merely imperfect measures
of threat (Richardson, 1960: 61). The psychological element is also derived and
discussed in more detail in Richardson's other work on 'war moods' (Richardson,
1948). Threat perception is a central element of most explanations for the arms
race-war connection (Sample, 2012). Nevertheless, the psychology of arms races
and their connection to war are often assumed or discussed, but not frequently
tested in experimental or empirical studies (see Jervis, 1978b; Kydd, 2000).

Most promising is Richardson's response to his critic that x is not merely a term
representing threat, but 'threats minus co-operation' (Richardson, 1960: 61, quote
marks in original). Often forgotten in applications of his model, this is a significant
qualification to models that seemed to portend runaway arms races. This suggests
that there can be substantial limits on arms races stemming from cooperation
between the competitors. As noted above, trade is used as a surrogate indicator for
such cooperation. Rivalry is treated as a possible accelerator for arms races.
Similarly, Senese & Vasquez (2008) list arms races as one of the 'steps to war'
along with other war-promoting conditions. In both cases, however, scholars add
other factors that make the escalatory effects of arms races worse rather than taking
the lead of Richardson and incorporating factors that mitigate those impacts.

The above suggestion that arms races are not deterministic is segue to some of
Richardson's other theorizing about their connection to war. Some of this is
inductively derived from an analysis of cases (see next section). *De facto*, he makes
a distinction between 'stable' and 'unstable' arms races, with the latter to be
regarded as dangerous (Richardson, 1960: 74–76). Richardson makes an important
distinction between the 'velocity' and the 'acceleration' of an arms race
(Richardson, 1960: 63). The former refers to the speed of the race, and can be
signified by the yearly rate of arms increases. The latter is the degree to which that
speed increases over time. Roughly, unstable arms races, and therefore ones likely
to end in war, are those in which the acceleration rate is increasing. In contrast,
stable arms races are those that show a slowing of the rate of increase and ones that
might promote a balance of power between the two enemies. Contrary to what
Richardson suggests about the importance of acceleration relative to velocity, more
contemporary scholars focus on yearly increases in military expenditures – speed –
and do not pay attention to whether this is increasing over time or not. This gap in
the literature suggests possible future lines of inquiry.

4.2.3   Empirical Findings

Although Richardson's work is more about mathematical modelling than empirical
testing, he does present some findings specifically on the arms race-war connection.
This is done by reference to his reading of historians' assessments, his own
quantitative analyses, and a series of case studies.

To his credit, Richardson was not an ahistorical modeler, but someone deeply steeped in actual cases with a good command of history. Accordingly, it is not surprising that he delved into the conclusions of historians as a way of framing his own investigations. Richardson notes that historians mention arms races for only 10 of 84 wars in the period 1820–1929; World War I is the most notable (Richardson, 1960: 70). He suggests that historians might have missed some arms races, and in any case they would not normally pay attention to statistics and in particular whether the acceleration was increasing or not (Richardson, 1960: 70); the latter is what he regards as a core component of dangerous arms races.

Richardson's own empirical analysis at the outset of the book suggests looking more carefully at different kinds of arms races (Richardson, 1960: 2ff). The domain of his examination is 15 countries in World War I. He examines the opposite of the *si vis pacem, para bellum* ('if you want peace, prepare for war') dictum. This is akin to the inverse of a deterrence argument, something that he later suggested might have the opposite effect: 'The evidence shows what was designed as a deterrent sometimes acted instead as an irritant' (Richardson, 1960: 55). He also looks at whether increased arms spending results in greater 'insurance' for the states involved, namely that it reduces the suffering should war break out. One of Richardson's dependent variable is total deaths (civilian and military) in the war. Thus, war onset and war severity for individual countries are muddled. He conducts several simple statistical tests between arms levels and outcomes (product moment correlations, Fisher's test) and finds no significant association between arms races and war, and with war suffering (Richardson, 1960: 8–11). Nevertheless, this initial exploration looks only at expenditures three years prior to a war (Richardson, 1960: 7), an unnecessary limitation that complicates the ability of the analyst to distinguish between spending in anticipation of war and spending that is causal in precipitating war; such a narrow window also is insufficient to detect acceleration (or lack thereof).

More detailed examinations are found later in the book, although again the focus is more on the dynamics of the arms races than their outcomes (Richardson, 1960: 7, Chs VI, VII). Examining some national interactions prior to World War I, Richardson considers how arms races produced (or not) stable equilibriums, namely balances of power. These get closer to testing his arguments about acceleration and (un)stable arms races. Yet the results are decidedly mixed, with Richardson expressing equivocation or ignorance as to whether arms races produce clear effects, with some cases fitting the argument and others not (see conclusions in Richardson, 1960: 7 and 76). Moving to the European context from 1908–14, there is greater consideration given to a longer time series for arms race data, although the author does aggregate spending within the two major alliances. It is difficult to discern precise conclusions about war, but the movement away from a point of balance and the positive instability coefficients suggest an arms race of a dangerous type. This is not surprising in that there is consensus even among arms race-war skeptics that World War I was preceded by an arms race. Richardson does conclude that there was equilibrium among European powers in 1907 and 1908, and had the arms race been muted by a small amount, the First World War might have been avoided (Richardson, 1960: 109–110). This counter-factual claim is one that is unlikely to be accepted by subsequent historians and international relations scholars.

More interesting is the examination of the relative impact of cooperation (measured as trade) vis-à-vis arms races. Trade between rival alliance blocs was increasing prior to World War I. There was also a trade agreement between Russia and Germany prior to World War II. Thus, the pacifying effects of trade are an important factor to consider. After considering trade patterns over many years, Richardson initially and optimistically notes that falling trade from 1929–33 was associated with a consistent high value of the instability coefficient, signaling danger (Richardson, 1960: 225). Nevertheless, Richardson is skeptical about the magnitude of the impact of trade (cooperation) on the negative consequences of arms races, concluding that '… to quell the subsequent arms race by reducing the instability coefficient to zero would have required, if proportional, an increase in international trade two and a half times more rapid than the exceptional fall' (Richardson, 1960: 225).

From today's research standards, it is easy to critique Richardson's empirical analyses. Obviously, he had a tendency to select on the dependent variable, primarily focusing on expenditures before wars, and in particular the two world wars; his analysis also looked principally at subsequent participants in those wars rather than giving equal attention to those that stayed out. Subsequent empirical research does a better job of this in that it includes a number of 'no war' as well as 'no arms race' cases. It also expands the empirical domain away from the world wars, although to give Richardson credit, he does a careful examination of the 19th century as well. Given Richardson's death in 1953, he was unable to examine the impact of nuclear weapons on arms races and the outbreak of war. At least one set of findings suggests that these might be a factor that mitigates any instability that arms races produce (Sample, 2012).

Richardson is to be commended for factoring in the power distribution and cooperation (trade) into his assessment about the dangers of arms races; defense burdens are also given consideration. The best subsequent research, specifically the 'steps to war' model (Senese & Vasquez, 2008), has also developed a multivariate and interactive approach. That work has also looked at defense burdens and power distribution gaps, but considered territorial disputes as well (see also Sample, 2012). That some arms races are dangerous and others not in a probabilistic fashion is consistent with Richardson's formulations, although later work is more developed theoretically. The importance of acceleration and the psychological impact of arms races, raised by Richardson, however, remain understudied.

References

Diehl, Paul F (1983) Arms races and escalation: A closer look. *Journal of Peace Research* 20(3): 205–212.

Diehl, Paul F (1985) Arms races to war: Testing some empirical linkages. *Sociological Quarterly* 26(3): 331–349.

Goertz, Gary; Bradford Jones & Paul F Diehl (2005) Maintenance processes in international rivalries. *Journal of Conflict Resolution* 49(5): 742–769.

Houweling, Henk W & Jan Geert Siccama (1981) The arms race-war relationship: Why serious disputes matter. *Arms Control* 2(2): 157–197.

Huntington, Samuel P (1958) Arms races: Prerequisites and results. *Public Policy* 8(1): 41–86.

Jervis, Robert (1978a) Cooperation under the security dilemma. *World Politics* 30(2): 167–174.

Jervis, Robert (1978b) *Perception and Misperception in International Politics*. Princeton, NJ: Princeton University Press.

Kydd, Andrew (2000) Arms races and arms control: Modeling the hawk perspective. *American Journal of Political Science* 44(2): 228–244.

Leeds, Brett Ashley (2003) Alliance reliability in times of war: Explaining state decisions to violate treaties. *International Organization* 57(4): 801–827.

Ray, James Lee (1990) Friends as foes: International conflict and wars between formal allies. In: Charles Gochman & Alan Ned Sabrosky (eds) *Prisoners of War: Nation-States in the Modern Era*. Lexington, KY: Lexington Books, 73–91.

Richardson, Lewis Fry (1948) War moods: I. *Psychometrika* 13(1): 147–174.

Richardson, Lewis Fry (1960) *Arms and Insecurity: A Mathematical Study of the Causes and Origins of War*. Pittsburgh, PA: Boxwood.

Rider, Toby; Michael Findley & Paul F Diehl (2011) Just part of the game?: Arms races, rivalry, and war. *Journal of Peace Research*, 48(1): 85–100.

Sample, Susan (2012) Arms Races: A cause or a symptom? In: John Vasquez (ed) *What Do We Know about War?* Lanham. MD: Rowman & Littlefield, 111–138.

Schneider, Gerald (2010) Economics and conflict: Moving beyond conjectures and correlations. In: Robert Denemark (ed) *International Studies Encyclopedia*. London: Wiley-Blackwell.

Senese, Paul & John Vasquez (2008) *The Steps to War: An Empirical Study*. Princeton, NJ: Princeton University Press.

Smith, Ron P (2020) The influence of the Richardson arms race model. Ch. 3 in this volume.

Wallace, Michael D (1979) Arms races and escalation: Some new evidence. *Journal of Conflict Resolution* 23(1): 3–16.

Zinnes, Dina (1980) Three puzzles in search of a researcher. *International Studies Quarterly* 24(3): 315–342.

Paul F. Diehl, b. 1958, Ph.D. in Political Science (University of Michigan, 1983); Ashbel Smith Professor of Political Science, Associate Provost, and Director, Center for Teaching and Learning, University of Texas at Dallas (2015–); most recent book *The Puzzle of Peace: The Evolution of Peace in the International System,* with Gary Goertz and Alexandru Balas (Oxford University Press, 2016), pdiehl@utdallas.edu.

Chapter 5
Richardson and the Study of Dynamic Conflict Processes

Kelly M. Kadera, Mark Crescenzi and Dina A. Zinnes

Abstract Lewis Fry Richardson made foundational contributions to the study of international relations. In this chapter, we examine his agenda-setting impact on the study of dynamics, time, and processes, especially conflict processes. We highlight the presence of Richardsonian dynamics in various formal and empirical models of peace and conflict. In so doing we emphasize the role of feedback and interactions in Richardson's models as well as other dynamic models, game-theoretic models, evolutionary game-theory and agent-based models, and quantitative empirical analyses. To show how these early foundations still inform research today, we then demonstrate how current research leverages dynamics to yield important insights concerning the origins and evolution of conflict such as when to expect norms of reciprocity to be present and to enable peace or exacerbate violence.

5.1 Introduction

Most important research questions in international politics concern processes with implicit time elements: how do crises evolve, when do hostile interactions become violent, how do shifts in trade affect alliances, when do regimes change? Answers inevitably require an explanation of how and why a process produces change over time. The pathbreaking work by Richardson (1960) helped us establish theories and methods for understanding time's role in IR. While Richardson was not the first to introduce the concept of time in international politics – see, e.g., the contemporaneous book by Sorokin (1957) on social and cultural change, which he linked to the emergence of war and the earlier work by Lanchester (1916) on attrition in combat[1] – he provided a unique and valuable perspective by linking it to the broader notion of dynamics.

---

[1]For a comparison of the models of Lanchester and Richardson, see MacKay (2020, in this volume).

N. P. Gleditsch (ed.), *Lewis Fry Richardson: His Intellectual Legacy and Influence in the Social Sciences*, Pioneers in Arts, Humanities, Science, Engineering, Practice 27, https://doi.org/10.1007/978-3-030-31589-4_5

Richardson's contributions are best seen through a comparison across various research traditions: game-theoretic models, evolutionary games and agent-based models (ABMs), differential equations (DEQ) models, and statistical analyses of time-structured data. We begin with a discussion of the concept of dynamics and then explore various approaches to time using illustrations from the literature. The differences across approaches highlight the significance of Richardson's contribution and suggest how scholars might incorporate it in future research.

5.2 Richardsonian Dynamics: Time, Feedback, and Interactions

The term *dynamics* is ubiquitous in the IR literature, and is often employed loosely to mean something along the lines of 'complicated and interesting.' *Webster's New Collegiate Dictionary* (1973) defines dynamics as 'the pattern of change or growth of an object or phenomenon'. Other definitions invoke terms like *variation*, *forces*, and *continuous change*. Time and process play essential roles. Change happens over time, according to some force acting on the phenomenon of interest. The word *dynamic* shares the same root as the Greek word *dyne*, which refers to 'the unit of force that would give a free mass of one gram an acceleration of one centimeter *per second per second*' (Webster's, 1973). As *dyne* is defined in terms of acceleration, its meaning is inherently tied to change over time.

Causal stories – arguments about what causes a particular variable to shift and take on different values – necessarily take place over time. Richardson's formulation of time, however, was more explicit. Although he sometimes used difference equations, in which a variable's value at time t is a function of that variable (or others) at previous times (e.g., $t - 1$ or $t - 2$), his canonical work used DEQs, where forces (represented by first or higher order derivatives) cause variables to increase or decrease over time.[2]

Richardson's models not only portrayed an over-time dynamic process, they included *feedback* and were *interactive*. In his classic arms-race model, each nation's current level of arms has a negative impact (feedback) on its own subsequent rate of armament, decreasing the rate of change in armaments (see Smith, 2020, in this volume, for Richardson's full arms-race model). At the same time, the feedback mechanism is interactive. A rival nation's increasing level of armaments positively impacts one's own calculations, increasing a state's rate of change in armaments.

---

[2]Brown (2007) thoroughly discusses the differences and similarities of difference equation models and DEQ models, paying particular attention to the types of trajectories each produces.

5.3   Dynamics in Formal Theory Traditions

Formal modeling traditions in IR – game theory, evolutionary games and ABMs, and DEQ models (which directly follow the Richardson tradition) – vary in how they feature dynamic elements. Here we sort out how each incorporates dynamics in its core characteristics and how those characteristics, or their absence, inform each tradition's understanding of global politics.

5.3.1   Game Theory

Game-theoretic models are commonly used in IR, and consist of players, actions (or strategies) they can choose, and payoffs associated with the outcomes produced by all possible combinations of actors' actions. Early game-theoretic models did not incorporate notions of time or process into any of these features of a game. In these simple, single shot, complete information games, players simultaneously choose their strategies and the outcomes, or Nash equilibria (NE), are immediately determined. Game-theoretic models have become considerably more complicated over time, though the basic structure remains.

Analyses of these models focus on finding equilibria, which are given by a list of the actions (strategies) each player would choose such that none would be better off by unilaterally changing to a different action (strategy). A sense of stickiness underlies the NE solution concept. Because no player has the incentive to unilaterally defect, a NE is inherently, or by its very design, stable. Osborne's description of a strategic form game highlights the absence of temporal features:

> Time is absent from the model. The idea is that each player chooses her action once and for all, and the players choose their actions 'simultaneously' in the sense that no player is informed, when she chooses her action, of the action chosen by any other player. … Nevertheless, an action may involve activities that extend over time, and may take into account an unlimited number of contingencies … However, the fact that time is absent from the model means that when analyzing a situation as a strategic game, we abstract from the complications that may arise if a player is allowed to change her plan as events unfold: we assume that actions are chosen once and for all (Osborne, 2004: 14).

As such, scholars anticipate that players identify and immediately implement strategies congruent with the NE, and that such equilibria are stable. Without time, feedback is missing. However, interaction resides in the interdependent strategies that structure the game.

As game theory evolved, scholars incorporated temporal features. Most notably, modelers sequenced players' actions and introduced incomplete information. An actor could now observe what another chose in a previous move, enhancing the sense of feedback and interaction. Innovations such as the Subgame Perfect Equilibrium (SPE) concept eliminates equilibria where incredible threats are made in sequential play (e.g., Fight if the other player Backs Down). Stability still exists, as SPE are a subset of the NE and thus invoke the same stickiness inherent in the original approach.

Sequential play conveys a sense of progression over time. An SPE found through backwards induction can easily be represented by a highlighted path from first action to last (starting from the last move and inferred up the extensive form tree to the beginning move). Sequential games with perfect information yield SPE characterized by a complete list of plays for what each player will choose contingent on what state of the game they are at (e.g., what all the players have chosen up until that point in time). These choices are discrete in time and each player's strategy can be written down as an entire plan of play (e.g., Fight if the other Challenges, Accept if the other Offers to Negotiate). As such, an SPE is completely determined at the outset and we can find the SPE from a collapsed, simultaneous play version of the game in strategic form, though we will also get some equilibria that are incredible. Thus, time provides some new information and helps us eliminate some unmeaningful equilibria, but the model's strategic insights remain rather static.

Games with incomplete information provide more insight into uncertainties of strategic decision making, but their central solution concept, the Perfect Bayesian Equilibrium (PBE) retains the stickiness of the NE and SPE. The PBE 'stipulates that all strategies are sequentially rational and consistent with beliefs that are updated according to Bayes's rule wherever possible' (Thomas, Reed & Wolford, 2016: 483). Players no longer know the payoffs of others, but instead assign probabilities to different payoff orderings. For example, Thomas et al. (2016) model rebel demands on governments, with war as one potential outcome. Rebels, R, do not know whether the government, G, can effectively fight or not. In turn, R do not know G's preferences for fighting versus yielding. Finding the PBE involves identification of 'the range of player-types of G that reject some proposal x' (Thomas et al., 2016: 506), R's choice of x such that it 'never backs down' in the final stage of the game. Even though R only knows G's type (or preferences), with some probability, and can update beliefs along the way, it decides at the outset what to do. Accordingly, 'In equilibrium, R always chooses a demand for which it will refuse to back down after rejection, making credible its threat to fight by increasing the range of government types that reject' (Thomas et al., 2016: 484). From the start, each player chooses a strategy, or a complete way to play the game. Players instantly reach and stay at the PBE. Players' immediate movement to the equilibrium and its stickiness characterize prominent and common games and solution concepts in IR research.[3]

Bargaining models with iteration incorporate time more explicitly (e.g., Fearon, 1998). In such models, players would both rather reach an agreement than to continue fighting but prefer different agreements. Players incur costs from fighting, and the game's features allow us to ask whether and when players will strike a

---

[3]More advanced games incorporate features such as errors in strategy, and third players, and use more complex solution concepts.

bargain, what it will look like, and what factors determine the answers to those questions. For example, Best & Bapat's (2018) model reveals that insurgent leaders reject government offers when internal divisions threaten to undermine insurgent cohesiveness. In such models, time and timing matters more explicitly, but feedback is still elusive.

5.3.2 Evolutionary Games and Agent-Based Models

Evolutionary games explicitly incorporate time into game structures but relax classic game theory assumptions (such as rationality) and forfeit some strategic insights. Axelrod & Hamilton (1981) develop PD tournaments in which actors adhere to one among several alternative strategies over the course of repeated plays. They notably demonstrate that cooperation can evolve because Tit-for-Tat (TFT), a 'nice' strategy that defects only if the opponent does so first, is evolutionarily stable. That is, when it dominates a population, TFT can survive invasion by mutant strategies. Analyses of evolutionary games focus on identifying evolutionarily stable strategies (ESS), but they differ from classic game theory strategies because they do not derive from actors' preferences; instead strategies are predetermined or inherited. When introducing replication dynamics, modelers also examine individuals' fitness for survival and temporal trends in the population of strategies.

Time plays at least two important roles in evolutionary games. First, results of past interactions feed back into the algorithm for an actor's current action. Second, patterns in the way (sub)populations of strategies rise and decline tell us whether and when we should empirically observe behavior like cooperation. Although few IR scholars use evolutionary games (but see, e.g., Johnson & Toft (2013/14) on fluctuations in territorial conflict), many leverage the concepts in exploring cooperation mechanisms such as norm change (Finnemore & Sikkink, 1998) or designing institutions for managing common pool resources (Ostrom, 1990).

Agent-based models similarly sacrifice traditional features of strategic interaction to capture how agent interactions produce temporal patterns at the aggregate level. ABMs add a social network to agent interactions: a grid system represents that network, agents are neighbors or distances separate them, and interactions depend on their closeness. Simulations reveal how populations of types of agents using various rules evolve, equilibria may not result, and may not be sticky. Scholars use ABMs to understand phenomena such as how coercion and emulation produce distinct temporal and geographic patterns in norm adoption (Ring, 2014) and the interplay between state formation and the severity of war (Cederman, 2003).[4]

---

[4]Cederman's (2003) agent-based model draws inspiration from Richardson's empirical work on power laws and wars. Cf. also Clauset (2020) and Spagat & van Weezel (2020), both in this volume.

5.3.3   Differential Equations Models

In contrast to classic game theory, but in concert with the spirit of evolutionary games and ABMs, Richardson and other scholars using DEQ models explicitly incorporate time, investigate more varied dynamic processes, and analyze their models with less focus on the equilibria themselves. Equilibria may or may not represent predictions of the model, but when they do, the issues concern stability. These distinctions translate into a greater focus on temporal features that characterize dynamic processes.

DEQ models specify at least one variable's values as a function of time. Typically, and in all the examples discussed in this section, a DEQ model uses more than one time-dependent variable, and the modeler builds a system of (first order) equations specifying how the change over time in one variable is driven by its own current value, other variables' current values, other variables' first (or higher) order derivatives, parameters, and the interactions of all of these elements. The resulting functions represent a variety of interdependencies and feedback loops.

A DEQ model's equilibrium is defined by the values of all the variables such that their first derivatives are zero. For example, in Richardson's arms race model, the equilibria are found by finding the value of x, nation X's level of armaments, and the value of y, nation Y's level of armaments, when $dx/dt = 0$ and $dy/dt = 0$. The no-change feature introduced by setting the first derivatives at zero seems, at first blush, to have some resemblance to a NE in game theory. However, not all equilibria are stable. That is, when a system in equilibrium is disturbed (e.g., shocked by an economic downturn that suddenly decreases one state's military spending), it might return to the equilibrium, and it might not. In the former case, we say the equilibrium is stable.[5] In the latter, we say it is unstable. Stability is assessed by analyzing the behavior of the trajectories, or paths over time, near an equilibrium.

Trajectories are paths produced by interdependencies and feedback. Any system has an infinite number of trajectories, but they can be grouped into different types: some spiraling outward, indefinitely escalating upward (as in the arms race case of a Richardson model) or downward (as in the 'love race' case of the Richardson model); settling into stable equilibria (as in the case where the joint defense burden outweighs the joint threats in the Richardson model, so that the arms race ends); endlessly cycling but never reaching equilibria (as in population biology's predator-prey model or some of the models of demographic, fiscal, and elite models of state expansion and collapse in Turchin, 2003); approaching the equilibrium from some directions but move away from it in others, forming a saddle shape around the equilibrium (as in some versions of the Muncaster & Zinnes, 1990, systemic hostility model); and so forth. Each type of trajectory is a little story about how the world might unfold, and savvy scholars can translate these types into substantively relevant vignettes about the causal process they are investigating. Multiple types of trajectories means multiple possibilities. In principle, once you

---

[5]An analogy in empirical time-series research would be the notion of mean-reversion.

know the initial conditions, the exact trajectory that the system takes is known, and assuming no exogenous shocks, the system is indeed deterministic.

Consider, for example, Kadera & Morey (2008), which examines how the trade-offs of fighting and investing yield different outcomes under three types of competitions between states: peacetime rivalry, counter-industrial wars, and counter-force wars. They model the first derivative with respect to time of four variables: nation I's military spending (m_i), nation I's level of resources (r_i), nation J's military spending (m_j), nation J's level of resources (r_j). Each is a function of at least two of the others, and the exact functional form depends on the kind of competition I and J engage in. Their analysis of the model yields graphs of the four variables over time. The authors notice an interesting pattern over the course of counterforce war trajectories: I's and J's military expenditures flatten out and stay equidistant from each other, locking in a particular level for m_i and for m_j, as well as for their difference, at all points in time. Kadera & Morey label this type of trajectory a *stalemate*. They conclude that in counterforce wars, 'Both states are ... eventually able to replace the military power that is being destroyed from the fighting without significantly harming their economies. The traditional war trajectories demonstrate that each state is capable of continuing the conflict indefinitely and neither is able to gain an advantage over the other' (Kadera & Morey, 2008: 167).

Differential equations models may not have equilibria, or current analytic methods may not allow the researcher to determine their explicit formulation. But the absence of equilibria does not negate the fact that the model can provide other, equally important insights such as trajectory behavior. Leveraging explicit equilibria values, Lee, Muncaster & Zinnes (1994) identify types of triadic friend and enemy relationship structures that persist or evolve into different structures, and Toft & Zhukov (2012) yield the conditions separating cases in which insurgency spreads across political subunits from those in which government coercion offsets transmission. Using simulations rather than explicit solutions or analytic identification of equilibria, Kadera, Crescenzi & Shannon (2003) show how a strong global democratic community insulates nascent democracies from the autocratization effects of war & Morey (2011) demonstrates how sudden upswings in fatalities shock a rivalry into termination by eroding support for continued hostilities.

Thus, two important points should be noted in comparing game theoretic and DEQ models. First, as Osborn (2004: 25) notes, 'Nash's theory concerns only equilibria; it has nothing to say about the path players' choices will take on the way to an equilibrium'. While game theoretic models reveal interesting interdependencies of decision making that are embedded in the final choices actors make, such as a rebel group making large demands that they know will be rejected by a government, in order to demonstrate their credibility (Thomas et al., 2016), DEQ models feature feedback that tells us about the type of route players or the system take and the speed of their travel. Second, the scholarly utility of a game theoretic model depends on whether it has few, identifiable equilibria; while the utility of a DEQ model relies less on the number or identifiability of equilibria and more on whether the trajectory behavior is substantively informative and interesting. In sum, if DEQ models represent what happens when actors do not 'stop to think', a famous characterization by Richardson, then game theoretic models represent what happens when actors do *nothing but* think.

5.4   Dynamic Processes in Research Design and Empirical Analysis

Although Richardson's work is most notable in its contribution to the modeling of international conflict, echoes of his approach can also be seen in quantitative empirical research. Action-reaction dynamics, for example, characterize early event coding and data. At the same time, scholars were grappling with the research design challenges of constructing statistical counterparts to Richardson arms race models (Schrodt, 1978). This early work on conflict processes continues to influence research today (e.g., Brandt et al., 2019) as is shown below in a brief overview.

5.4.1   Early Work

Early Richardsonian empirical work falls into two categories: (1) research focusing directly on Richardson's arms race model and (2) efforts to collect data on international events. Not long after Zinnes and others began working on theoretical models drawing from Richardson's work, empirical analyses of Richardson's work also began to emerge. Majeski & Jones (1981), for example, revise and operationalize Richardson's equations in an attempt to better understand dyadic arms races, finding no evidence of an action-reaction process. Ward (1984) also leverages Richardson's equations when analyzing the arms race between the US and the USSR. Ward highlights the problem of simplistic operationalizations of the equations, particularly a confusion between arms stockpiles and current military budgets. He refocuses the work on perceived differences in stockpiles as a motivator for action and reaction, and in so doing finds that the US and the USSR participate 'in a reaction process in which each stimulates the other to spend more on their military establishments' (Ward, 1984: 202).

While the empirical work during this time is not explicitly driven by Richardson, it is possible to identify his influence more abstractly in the genesis of the events data sets begun in the 1970s. The *World Events Interaction Survey* (WEIS) codebook, for example, hints at dynamic processes when it refers to how the data 'reflect the flow of action and response between countries' (McClelland, 1978: 1), even though McClelland designed the WEIS project to merely catalogue the chronology of international events.

A few years later, the Conflict and Peace Data Bank (COPDAB) (Azar, 1980), was motivated largely by new perspectives on IR research, as laid out in Zinnes (1976). Whereas McClelland resisted scaling events, the COPDAB project ranked events along a single dimension of conflict and cooperation. These two event data sets (WEIS and COPDAB) stimulated a large body of scholarship. Although scholars identified serious challenges to statistically modeling and fitting Richardson arms race models (Schrodt, 1978), explorations of the broader notion of action-reaction dynamics developed by Zinnes (1980) and others (e.g., Gillespie et al., 1977) were well served by event data.

Goldstein (1992) used an expert survey to create a scale to migrate WEIS categories to a conflict-cooperation scale, launching a new wave of event-data research. This innovation enabled Goldstein & Pevehouse (1997, see also Pevehouse & Goldstein, 1999), for example, to examine dyadic and triadic reciprocity dynamics in Bosnia and Serbia, and Schrodt & Gerner (1997) to examine phase shift dynamics in the Middle East. At the same time, Gerner & Schrodt (1996) developed the first machine-coded event data generation processes, known as the Kansas Event Data System (KEDS), building on the infrastructure created by McClelland two decades prior.

5.4.2 Current Empirical Research: An Illustration

Quantitative empirical studies of conflict dynamics continue to develop. One contemporary empirical approach is seen in new research by Brandt, Freeman, Lin & Schrodt (2019). They apply time series models to analyze the conflict interaction patterns between dyads (pairs of states or actors) over time to pinpoint phases of reactivity versus phases of independent action. The focus is on three independent streams of conflict interactions between: Israel & Palestinian groups, China & Taiwan, and India & Pakistan. Their search for phase shifts reveals that all three sets of interactions contain similar shifts between high and low-entropy, where entropy refers to the volatility and variance of actions and reactions within each dyad.[6]

Additionally, the authors find patterns that indicate the use of norms of reciprocity, wherein states react in kind to cooperative or conflictual events. Norms of reciprocation are useful in producing interactions conducive to peace and stability. The major substantive finding of the paper is that norms of reciprocity are more likely to govern behavior in low entropy phases. Thus, high-volatility phases create conditions under which leaders have a difficult time conveying and perceiving information about their intentions and the intentions of others. Reputations become harder to establish, signals are more difficult to send, and the problems relating to the analog of private information in bargaining environments become more pernicious.

Brandt et al. (2019) offer two important lessons: First, the explicit use of statistical methods to discern between different patterns of interactions over time could

---

[6]Based on our reading of Brandt et al. (2019), low-entropy is analogous to a stable equilibrium in a dynamic model, and high-entropy is analogous to an unstable equilibrium.

uncover the complex dynamics of phase shifts and therefore help scholars understand *when* Richardson-like action-reaction processes occur. As such, the research design matters. Second, they find that the differences in the risk of severe conflict substantially differ across high and low entropy phases. Failing to take these differences into account runs the risk of pooling action-reaction patterns across high and low entropy phases and arriving at incorrect conclusions regarding the risk of war.

5.5   Conclusions and Recommendations

Richardson made a foundational contribution to the study of conflict – defining dynamics as an interactive, feedback-based, temporal process. Differential equations models explore the dynamics produced by interactive feedback mechanisms that highlight a rich variety of over-time patterns. In contrast, the focus of game theory is on interactions and expected interactions in the form of actor strategies and substantively important equilibrium solutions. Game-theoretic models do not explicitly capture interactions over time. Evolutionary games and ABMs marry some features of game theory with more dynamic elements that yield meaningful temporal patterns. In empirical work, theoretical dynamic processes are embedded in statistical research designs, while event data analyses utilize dynamic assumptions embedded in econometric models.

All the approaches reviewed here contribute to the study of world politics and contain an important lesson: explicitly incorporating dynamics and feedback mechanisms into theory, research design, and analysis provides valuable insight into causal processes (see Crescenzi & Kadera, 2015). Our hope is that future scholars pay close attention to the role that dynamic processes can play in understanding international politics. Ignoring the impact of dynamics can mean the difference between finding or missing the answer to a research puzzle.

References

Axelrod, Robert & William D Hamilton (1981) The evolution of cooperation. *Science* 211(4489): 1390–1396.

Azar, Edward E (1980) The conflict and peace data bank (COPDAB) project. *Journal of Conflict Resolution* 24(1): 143–152.

Best, Rebecca & Nathan Bapat (2018) Bargaining with insurgencies in the shadow of infighting. *Journal of Global Security Studies* 3(1): 23–37.

Brandt, Patrick; John Freeman, Tse-min Lin & Philip Schrodt (2019) A Bayesian time series approach to the comparison of conflict dynamics. *Political Science Research and Methods*, in press.

Brown, Courtney (2007) *Differential Equations: A Modeling Approach.* Los Angeles, CA: Sage.

Cederman, Lars-Erik (2003) Modeling the size of wars: From billiard balls to sandpiles. *American Political Science Review* 97(1): 135–150.

Clauset, Aaron (2020) On the frequency and severity of interstate wars. Ch. 10 in this volume.

Crescenzi, Mark JC & Kelly M Kadera (2015) Built to last: Understanding the link between democracy and conflict in the international system. *International Studies Quarterly* 60(3): 565–572.

Fearon, James (1998) Bargaining enforcement and international cooperation. *International Organization* 52(2): 269–305.

Finnemore, Martha & Kathryn Sikkink (1998) International norm dynamics and political change. *International Organization* 52(4): 887–917.

Gerner, Deborah J & Philip A Schrodt (1996) The Kansas event data system: A beginner's guide with an application to the study of media fatigue in the Palestinian intifada. Poster presented at the Annual Meeting of the American Political Science Association, San Francisco, CA.

Gillespie, John V; Dina A Zinnes, GS Tahim, Philip A Schrodt & R Michael Rubison (1977) An optimal control model of arms races. *American Political Science Review* 71(1): 226–244.

Goldstein, Joshua S (1992) A conflict-cooperation scale for WEIS events data. *Journal of Conflict Resolution* 36(2): 369–385.

Goldstein, Joshua S & Jon C Pevehouse (1997) Reciprocity, bullying, and international cooperation: Time-series analysis of the Bosnia conflict. *American Political Science Review* 91(3): 515–529.

Johnson, Dominic DP & Monica Duffy Toft (2013/14) Grounds for war: The evolution of territorial conflict. *International Security* 38(3): 7–38.

Kadera, Kelly M; Mark Crescenzi & Megan Shannon (2003) Democratic survival, peace, and war in the international system. *American Journal of Political Science* 47(2): 234–247.

Kadera, Kelly M & Daniel S Morey (2008) The trade-offs of fighting and investing: A model of the evolution of war and peace. *Conflict Management and Peace Science* 25(2): 152–170.

Lanchester, Frederick W (1916) *Aircraft in Warfare: The Dawn of the Fourth Arm*. London: Constable.

Lee, Sung-Chull; Robert Muncaster & Dina Zinnes (1994) 'The friend of my enemy is my enemy': Modeling triadic internation relationships. *Synthese* 100(3): 333–358.

McClelland, Charles (1978) *World Event/Interaction Survey: (WEIS), 1966–1978*. Ann Arbor, MI: Inter-university Consortium for Political and Social Research.

MacKay, Niall (2020) When Lanchester met Richardson: The interaction of warfare with psychology. Ch. 9 in this volume.

Majeski, Stephen & David Jones (1981) Arms race modeling: Causality analysis and model specification. *Journal of Conflict Resolution* 25(2): 259–288.

Morey, Daniel (2011) When war brings peace: A dynamic model of the rivalry process. *American Journal of Political Science* 55(2): 263–275.

Muncaster, Robert G & Dina Zinnes (1990) Structure and hostility in international systems. *Journal of Theoretical Politics* 2(1): 31–58.

Osborne, Martin J (2004) *An Introduction to Game Theory*. New York: Oxford University Press.

Ostrom, Elinor (1990) *Governing the Commons: The Evolution of Institutions for Collective Action*. Cambridge: Cambridge University Press.

Pevehouse, Jon C & Joshua S Goldstein (1999) Serbian compliance or defiance in Kosovo? Statistical analysis and real-time predictions. *Journal of Conflict Resolution* 43(4): 538–546.

Richardson, Lewis Fry (1960) *Arms and Insecurity: A Mathematical Study of the Causes and Origins of War*. Pittsburgh, PA: Boxwood.

Ring, Jonathan Jacob (2014) *The Diffusion of Norms in the International System*. PhD Dissertation. University of Iowa.

Schrodt, Philip A (1978) Statistical problems associated with the Richardson arms race model. *Journal of Peace Science* 3(2): 159–172.

Schrodt, Philip A & Deborah J Gerner (1997) Empirical indicators of crisis phase in the Middle East, 1979–1995. *Journal of Conflict Resolution* 41(4): 529–552.

Smith, Ron P (2020) The influence of the Richardson arms race model. Ch. 3 in this volume.

Sorokin, Pitirim (1957) *Social and Cultural Dynamics: A Study of Change in Major Systems of Art, Truth, Ethics, Law and Social Relationships*. Reprinted, 1970. Boston, MA: Porter Sargent.

Spagat, Michael & Stijn van Weezel (2020) The decline of war since 1950: New evidence. Ch. 11 in this volume.

Thomas, Jakana L; William Reed & Scott Wolford (2016) The rebels' credibility dilemma. *International Organization* 70(3): 477–511.

Toft, Monica Duffy & Yuri M Zhukov (2012) Denial and punishment in the North Caucasus: Evaluating the effectiveness of coercive counter-insurgency. *Journal of Peace Research* 49(6): 785–800.

Turchin, Peter (2003) *Historical Dynamics: Why States Rise and Fall.* Princeton, NJ: Princeton University Press.

Ward, Michael D (1984) Modeling the USA-USSR arms race. *Simulation* 43(4): 196–203.

Webster's New Collegiate Dictionary (1973). Springfield, MA: Merriam.

Zinnes, Dina A (1976) *Contemporary Research in International Relations: A Perspective and a Critical Appraisal.* New York: Free Press.

Zinnes, Dina A (1980) Three puzzles in search of a researcher: Presidential address. *International Studies Quarterly* 24(3): 315–342.

Kelly M. Kadera, b. 1965, Ph.D. in Political Science (University of Illinois, 1993); Professor, University of Iowa (1993–); Vice-President, International Studies Association (2016–17); current main interests: dynamics of conflict and democracy, conflict environments, gender and scholarship in International Relations, kelly-kadera@uiowa.edu

Mark J. C. Crescenzi, b. 1970, Ph.D. in Political Science (University of Illinois, 2000); Professor, University of North Carolina at Chapel Hill (1999–); current main interests: reputation and learning in international politics, conflict environments and civil wars, dynamics of democracy, crescenzi@unc.edu.

Dina A. Zinnes, b. 1935, Ph.D. in Political Science (Stanford University, 1964); Professor of Political Science, Indiana University (1967–80), Merriam Professor of Political Science, University of Illinois (1980–2005), now emeritus. Founded and directed the Merriam Laboratory for Analytic Political Research (1986–2005). President of the International Studies Association (1980–81); Editor, American Political Science Review (1981–85), President, Peace Science Society (1989), zinnes@illinois.edu.

Chapter 6
Back to the Future: Richardson's Multilateral Arms Race Model

Michael D. Ward

Abstract Lewis Fry Richardson was a groundbreaking scholar, not only in modern meteorology but also in world affairs. His two major books, both published posthumously in 1960, were harbingers for what was to follow in scholarly international relations. In one, he collected wide-ranging, detailed quantitative information on disaggregated conflict processes in a variety of historical contexts. In the other, he showed two basic innovations for the social sciences: (a) the power of mathematics for understanding complex social systems and (b) the importance of understanding the interdependence of things that are typically studied separately. That latter insight is the focus of this chapter wherein I show that the Richardsonian insight on coupled behaviors leads to a network perspective on social interactions at the global scale. We call these coupled interactions *networks*. I trace the development of Richardson's thinking about coupled phenomena to the development of network thinking in the social sciences. I conclude with some recommendations for the arms race research program as applied to the current era.

6.1 Introduction

Based on his path-breaking work in meteorology (Richardson, 1922), Richardson imagined a future in which his models were implemented for real time forecasting. The illustration is shown in Fig. 6.1. He imagines a spherical structure which is mapped to a globe and painted correspondingly inside. People calculate the prediction equations corresponding to which part of the map they are assigned to.

The original version of this chapter was revised: The name "Gregory D. Hess" has been corrected to "George D. Hess" in Reference list. The correction to this chapter is available at https://doi.org/10.1007/978-3-030-31589-4_12

Thanks to Phil Schrodt five decades of inspiration and to Nils Petter Gleditsch for his stewardship.

Fig. 6.1 Richardson's visualization of a laboratory for real-time weather forecasting, based upon his mathematical models. *Source* Richardson (1922: 219)

He imagined 64,000 people – which he called calculators – would be required. A team would distribute the calculation of each part of his equation system, and it would be coordinated and monitored by supervisors. At the top of a pillar inside the sphere, a conductor is in charge of all teams. The musicians in this sense are individuals playing slide-rules and calculating machines.

Richardson's work in meteorology continues to be important to this day. Indeed, they have played a role in the discovery of global warming (NOAA National Weather Service, 2019; Weart, 2008, 2018) through their role in global circulation models of weather.

However, when Lewis Fry Richardson learned that his weather models could be useful in military applications, he quit working in this domain and reportedly destroyed his unpublished results. Richardson was a Quaker and would eventually spend part of the First World War as an ambulance driver (Section Sanitaire Anglaise 13) from 1916 to 1920 (Wilkinson, 1980). During this time, he wrote his first offering on the causes of war, but at the time there was nowhere to publish it. During his time in France, his location was under frequent bombardment and not only were the ambulances quite busy, they also had to run the gauntlet to deliver the wounded to medical attention. This reportedly had a large impact on Richardson.

When he turned from weather prediction, he focused on wars and conflicts, notably in two volumes. One of these concentrated on analyzing data (Richardson,

1960b) and the other focused on mathematical analysis of the insecurity of nations (Richardson, 1960a). In that latter effort, he drew upon his development of mathematical equations in meteorology and applied them to human behavior. His initial focus was on the arms race that preceded the First World War. In terms of his modeling, this is his best-known work.[1]

Although he was not the first person to use the term arms race, he was the first to formalize what it meant precisely. Richardson begins with a linear model for two nations, expressed as a pair of differential equations (1960b: 16, Eqs. 7 & 8):

$$dx/dt = ky - \alpha x + g \tag{1}$$
$$dy/dt = lx - \beta y + h \tag{2}$$

where x represents the military spending of one nation and y the military spending of its main rival. The drag on increasing military spending at ever higher levels is represented by the term $-\alpha x$ for the first country (aka country x) and $-\beta y$ for the second country (referred to as country y); g and h are historical constants reflecting the respective hostility of x and y toward each other.

In the 1960s when this idea was introduced into the study of politics and economics, it was very difficult (for social scientists) to solve differential equations. Even Richardson's initial attempts to use data to look at these equations were quite simplistic. Basically, everyone translated these into a set of difference equations

$$\Delta x = ky_t - \alpha x_t + gt \tag{3}$$
$$\Delta y = lx_t - \beta y_t + h_t \tag{4}$$

which could be dealt with via straightforward mathematical tools and could (if you squinted) be examined empirically via linear regression – possibly via a two-stage least squares estimator. Richardson analyzed this analytically in his volume and several scholars worked in this domain quite successfully. Hess (1995) provides a good overview.

During the Cold War, there was widespread empirical work on arms races, frequently using a statistical approach to estimating and validating the underlying action-reaction equations developed by Richardson in *Arms and Insecurity*.[2] Following Richardson's lead most of these studies looked at pairs of countries, or groups of countries aggregated into pairs (NATO versus Warsaw Pact). Readers will notice that most of these studies were conducted in the last century during the (first?) Cold War. Dissolution of the Soviet Union led to a subsequent scholarly focus on the putative peace dividend. In the meantime, global military spending has grown from about one trillion dollars per annum to a current total of around 1.7

---

[1] Although he made contributions to geography (e.g., the theory of compactness) (see Gleditsch & Weidmann, 2020, in this volume) and to other disciplines as well, such as criminology.

[2] For an extensive bibliography of such works, see Gleditsch & Njølstad (1990: 384ff). For references to more recent work, see Smith (2020) and Diehl (2020), both in this volume.

trillion dollars per annum (SIPRI, 2018). But a search for recent articles on arms races leads to a mountain of research on biological interactions from the microscopic to the species level, but little contemporary social science (Smith, 2020, in this volume).

6.2 Multilateral Arms Races

The two-nation arms race was actually the toy model through which Richardson introduced his basic ideas. He quickly moved beyond that, though most scholarly work has not. A sterling exception is Schrodt (1981), who focused on a multi-polar world with more nuanced distributions of armaments (and therefore spending).[3]

The basic structure of a multi-nation arms race, in Richardson's terms, much like weather systems is given as a system of ordinary differential equations as shown in Eq. 5, where x_i is the military spending for nation i, and $\kappa_{i,j}$ has the action-reaction coefficients off the diagonal, and the economic constraints on the diagonal and g_i portrays the hostility terms.

$$\frac{dx_i}{dt} = g_i + \sum_{j=1}^{j=n} k_{ij}x_j \quad \forall \quad i \in \{1, 2, 3, \ldots, n\} \tag{5}$$

This equation is the crux of the multilateral system of equations. Instead of x and y there is now a vector of countries stored in x_i, where i is an index of all the countries to be included (including the previous y from the bilateral case). κ is an $i \times i$ matrix. The off-diagonal elements collect the action-reaction terms, linking each i to each other i with a coefficient that conveys the reaction of a single country to each other countries' military spending. For example, in the first row of Table 6.1, there is a weak reactivity between Czechoslovakia and Germany (2). These effects are asymmetric as shown by the coefficient of 36 between Germany and Czechoslovakia. Germany is threatened by small changes whereas Germany is more reactive to changes in the military spending of Czechoslovakia. Collected on the diagonals are the economic constraints, wherein higher spending tends to dampen subsequent spending in the same country (thus the negative sign). Turning to the first element, there is a strong economic constraint in Czechoslovakia (-30) which constrains military spending increases. The country with the greatest economic constraint is Great Britain and Northern Ireland (-45). The overall hostility terms from the two-nation model are collected in the vector g for each country; these are not shown. In the end, this representation is equivalent to writing out a Richardson model for each country; what is unique is that it generalizes for any number of countries.

---

[3]There are a few exceptions to this generalization, notably Wallace (1979).

Table 6.1 The κ matrix from Richardson's study of the 1935 multi-nation arms race

	Czech	China	France	Germany	GBNI	Italy	Japan	Poland	USA	USSR
Czech	−30			2				1		
China		−30					12			18
France			−54	4		4				
Germany	36		36	−30	18			3		72
GBNI				4	−45	6	2			
Italy			18		36	−15				18
Japan		12					−30		36	36
Poland	9			3				−30		9
USA				2	6	2	4		−21	6
USSR		2		8	6	2	4	1		−30

Values are multiplied by 30 for presentation purposes, as in the original. The determinant (unscaled) of this matrix is −0.37. *Source* Richardson (1960a: 202, Table 51)

The κ coefficients were calculated by Richardson, but not necessarily using any statistical methodology. Instead, he coded them from subject matter expertise, where Richardson was the expert having closely followed current events in the news and radio reporting of the day. If the determinant of the κ_{ij} is not zero, the equilibrium conditions can be easily solved.

6.3   Military Spending

Richardson's initial study focused on military spending. His Table 1 (1960a: 6–7) portrays a variety of data about countries over the pre-war period from 1913–15, but he also collected information on the number of war dead and population. These data were taken from various almanacs, historical sources, and Parliamentary documents. Subsequent uses of the Richardson arms-race model largely focused on military spending data – often normalized by population or total governmental expenditures.

The data from his study of the multilateral arms race in 1935 (Richardson, 1960a: 202ff) are given in Table 6.1. These $\kappa_{i,j}$ are basically the linkages or reactivity of each nation to each other nation. A value of zero indicates that the two nations essentially disregard each other's military spending. Diagonals convey a negative value that reflects the economic constraints faced by each country.

Many have argued that the arms race leading to the onset of the First World War was actually a competition over military equipment, notably Navies (Lambelet et al., 1979), and my own position is that it really makes more sense to think of arms races in terms of stock-flow models (Ward, 1984a). However, for the purposes of this exposition, I focus on military expenditure data, eschewing the obligation of developing a coherent capital stock measure for military technologies in the 21st

Table 6.2 Military spending in thirteen countries, in million US $ 2016

Country	2016	2017
US	600,106	597,178
China	216,031	228,173
Russia	69,245	55,327
Saudi Arabia	63,673	69,521
France	57,358	56,287
India	56,638	59,757
UK	48,119	48,383
Japan	46,471	46,556
Germany	41,579	43,023
South Korea	36,934	37,560
Israel	14,783	15,501
North Korea	13,000	13,000
Pakistan	9,974	10,378

Century. Data have been taken from the Stockholm International Peace Research Institute (SIPRI) which has been the gold standard for military spending data for several decades.

Military spending in 2016 US constant dollars is from SIPRI military spending database (accessed 28.6.2018). For North Korea data are based on an estimate of one-third of the budget. GNP is estimated in the CIA *World Factbook* at 40 billion current dollars. Thus, I have estimated military spending to be about 13 billion dollars per annum.

Table 6.2 portrays military expenditure data for thirteen countries taken from the SIPRI database. These data illustrate that the US and China each have an order of magnitude more military spending than any other country, though current US expenditures are not as high as they were at the height of the so-called Cold War. The US current spends over twice what China spends on an annual basis, though factors of production vary widely between these two countries, as they do for many others in this list. It also serves to point out that a variety of activities undertaken by the military are accounted in different ways by different countries. Some countries do not include space exploration and satellite activities as military ones. Both intelligence and financial activities sometimes come under a military purview and other times do not. Thus, an exact comparison is hard to justify in a nuanced way. Sometimes keeping aging weapons systems is more expensive than buying new ones, but procurement may dependent upon legislative oversight and always takes time. As a result, military expenditures are a complicated mixture of procurement of new systems as well as parts and services for aging systems. Obviously, the cost for a battalion of infantry varies substantially across these countries, spanning Pakistan, North Korea, Israel, Saudi Arabia, and many western European countries. Suffice it to say, however, that the US and China are outliers in the distribution of military spending for contemporary countries.

The annual changes in these expenditures are not extreme. Saudi Arabia has the greatest growth in military spending, about 9% per annum, with China, India, and Israel close behind at about 5–6% growth per annum. All the other countries are within a percent or two of exhibiting a rate of change that is close to zero. Russia is negative, but not at any substantial level. I return to these data below.

6.4   Reactivity

Richardson had a complicated way of estimating the coefficients for his model. He describes his approach:

> In September 1938 the author made the assumptions shown in Table 45, remarking, with apologies to all concerned, that it was an act partaking not only of science but also of art or perhaps, alas, of caricature. For several years he had been attending to the profuse comment on the friendships and animosities of nations offered by various publications but especially by the British radio and the *Glasgow Herald*. This matrix was intended to be a general estimate and summary of such common information relating to the year 1935. (Richardson, 1960a: 193)

This can be considered as a network of interactions among these thirteen countries (ignoring for the moment the diagonals). This is shown in Fig. 6.2.

6.5   The System of Equations

Richardson never got to see his room full of weather calculators, nor to see how his application of ordinary differential equations might play out when examined with data. Solving the differential systems of equations by hand was possible analytically, but deriving real trajectories was a technology then yet to come. Fortunately, since then the science and engineering of dynamical systems has made enormous progress.

How much progress? In the course of a weekend eight decades after Richardson's work, I encoded Richardson's κ_{ij} matrix for 1938, and with military expenditure data taken from Richardson's own source (*The Literary Digest*, 23 February 1935: 42) was able to implement this 10-national differential equation system.[4] Figure 6.3 shows the trajectories of four major countries in this ten-country arms race: France, Great Britain and Northern Ireland, Soviet Union, and Germany. There is explosive growth in these trajectories. This shows, as claimed by Richardson (1960a: 12), that 'The equations are merely a description of what people would do if they did not stop to think.' Richardson was ambivalent

---

[4] I am unaware of anyone else having done this.

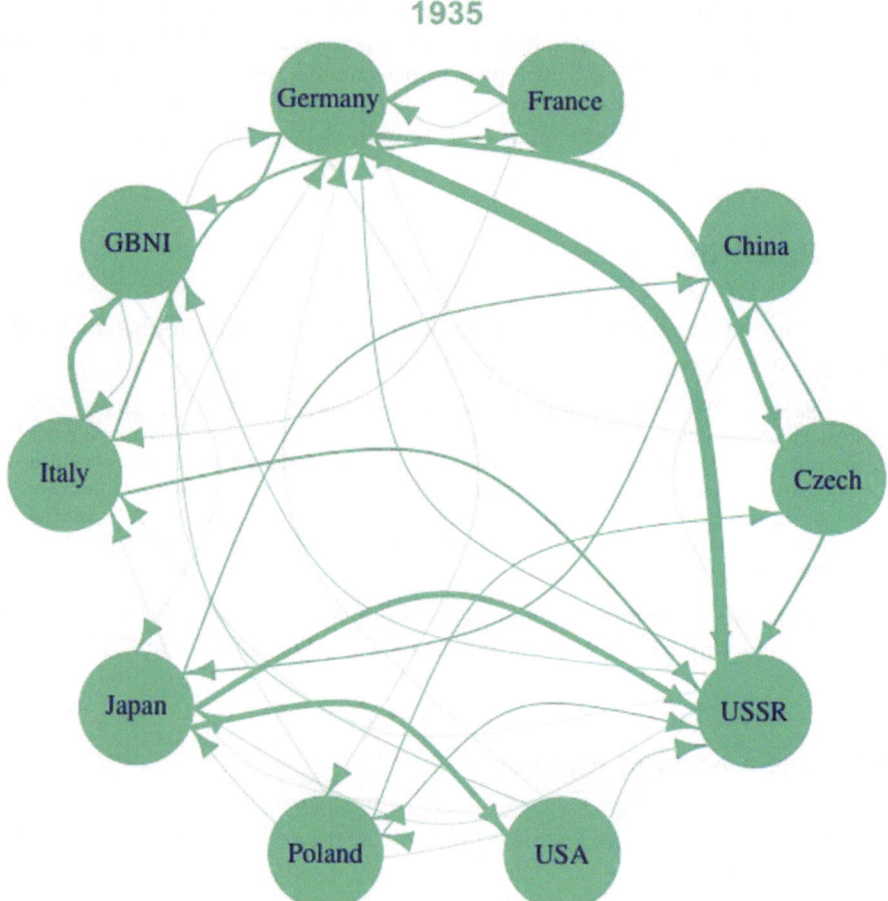

Fig. 6.2 Richardson's $\kappa_{i,j}$ visualized as a weighted network

about whether arms races brought about wars, but it should be obvious that this system was embroiled in the Second World War by September of 1939, and the trajectories of spending by these countries did, indeed, grow exponentially.

Richardson clearly underestimated the damping effect – in peace time at least – of high levels of spending on subsequent increases. This system probably overestimates the explosive nature of the arms expenditures because of their reactivity to spending in a large number of potential rivals (and allies). Several scholars have worked to point this out in empirical studies of arms races, beginning with Caspary (1967).

In addition, scholars have pointed out that it is not really all about military spending, but rather what the spending is buying that should be the central focus of competitive arms processes (Luterbacher, 1974; Ward, 1984, 1985). In economics, these are called stock-flow models. Richardson's differential equation system can

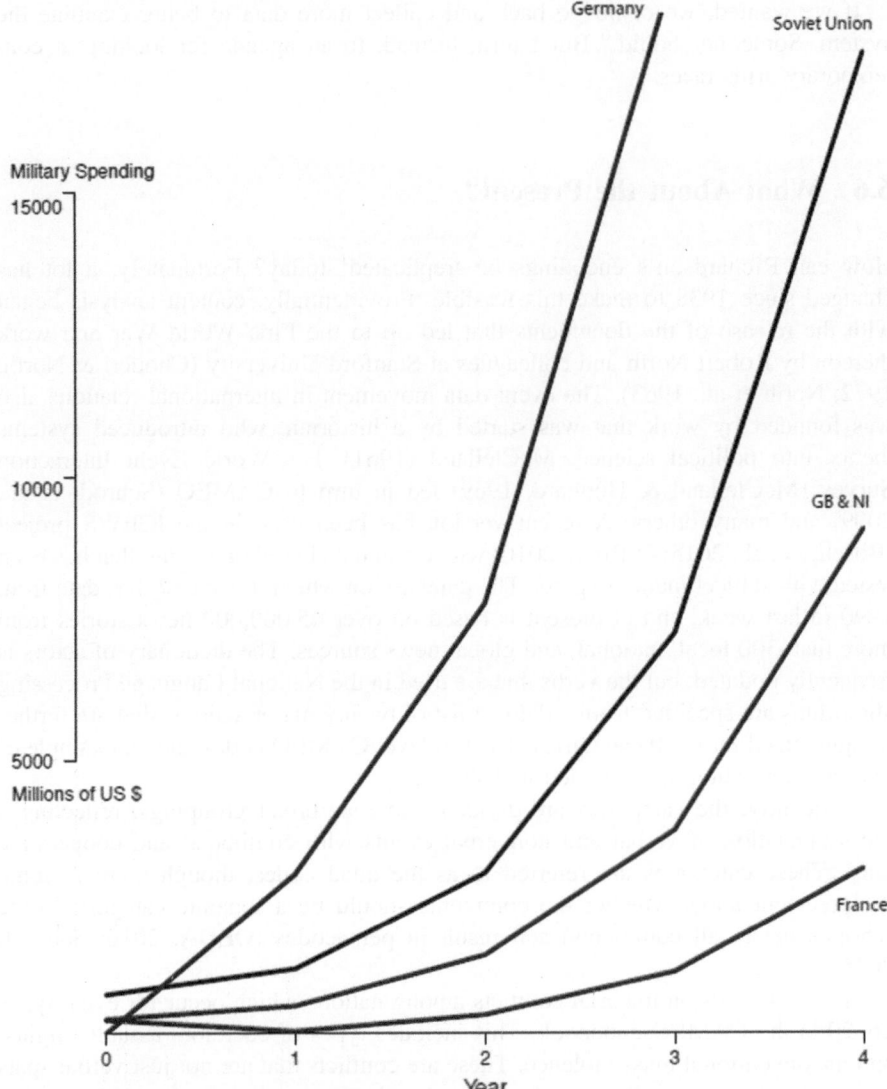

Fig. 6.3 Trajectories of four countries in Richardson's 10-nation arms race. *Source* Specified by his estimates of κ_{ij} in Richardson (1960a: 202)

easily handle both emendations – though collecting accurate and comparable data on stockpiles of weapons has proven challenging. Even the US Central Intelligence Agency had great difficulty of this in the 1970s and 1980s.[5]

---

[5]See the description of the Team B project, led by Richard Pipes (Pipes, 1986).

If we wanted, we could go back and collect more data to better examine the system. Someone should.[6] But I turn, instead, to an agenda for looking at contemporary arms races.

6.6 What About the Present?

How can Richardson's encodings be 'replicated' today? Fortunately, a lot has changed since 1938 to make this feasible. Providentially, content analysis began with the release of the documents that led up to the First World War and work thereon by Robert North and colleagues at Stanford University (Choucri & North, 1972; North et al., 1963). The event data movement in international relations also was founded by work that was started by a historian, who introduced systems theory into political science, McClelland (1961). His World Event Interaction Survey (McClelland & Hoggard, 1969) led in turn to CAMEO (Schrodt et al., 2009) and many others. A recent version has been used in the ICEWS project (Boschee et al., 2015; O'Brien, 2010) with event data based on coding that has been tested with subject matter experts. The database on which it is based, has data from 1990 to last week, and at present is based on over 45,000,000 news stories from more than 300 local, national, and global news sources. The dictionary of actors is frequently updated, but the verbs that are used in the National Language Processing algorithms are specified from a defined list of twenty major actions, that are further disaggregated into 360 categories. The top-level CAMEO codes and an example of one disaggregation are provided in Table 6.3.

Frequently, the categories are divided into four broad groupings, reflecting a cross tabulation of verbal and nonverbal events with conflictual and cooperative ones. These categories are referred to as the quad codes, though there is some disagreement about whether the comments should be a separate category (some scholars delete all comments) and result in pentacodes (OEDA, 2016; Schrodt, 2015).

Herein, I focus on material conflicts among nations which occur for event types 16–20 in the CAMEO codebook. This includes types of coercion, assaults, fights, and unconventional mass violence. These are conflicts that are not just verbal spats between nations, but rather are those that involve some use of material resources. This variable is typically called material conflict. I gather the counts of bilateral material conflict events from the ICEWS database. Herein, I use an annual aggregation, though any sensible temporal aggregation greater than a day is possible. The data for 2017 are shown in Table 6.4.

---

[6]This, as they say, is left to the reader as an exercise.

Table 6.3 The CAMEO Codes, and a disaggregation of category 03

Cameo Codes for Event Types

01 MAKE PUBLIC STATEMENT	11 DISAPPROVE
02 APPEAL	12 REJECT
03 EXPRESS INTENT TO COOPERATE	13 THREATEN
04 CONSULT	14 PROTEST
05 ENGAGE IN DIPLOMATIC COOPERATION	15 EXHIBIT FORCE POSTURE
06 ENGAGE IN MATERIAL COOPERATION	16 REDUCE RELATIONS
07 PROVIDE AID	17 COERCE
08 YIELD	18 ASSAULT
09 INVESTIGATE	19 FIGHT
10 DEMAND	20 USE UNCONVENTIONAL MASS VIOLENCE

```
03       EXPRESS INTENT TO COOPERATE
030          Express intent to cooperate, not specified elsewhere
031          Express intent to engage in material cooperation
0311            Express intent to cooperate economically
0312            Express intent to cooperate militarily
0313            Express intent to cooperate on judicial matters
0314            Express intent to cooperate on intelligence
032          Express intent to engage in diplomatic cooperation
033          Express intent to provide material aid, not specified elsewhere
0331            Express intent to provide economic aid
0332            Express intent to provide military aid
0333            Express intent to provide humanitarian aid
0334            Express intent to provide military protection or peacekeeping
```

Adapted from the CAMEO Codebook by the author, http://eventdata.parusanalytics.com/cameo.dir/CAMEO.09b6.pdf

Table 6.4 Annual counts of material conflict events sent by the country on the row toward the country on the column in 2017

	US	Ch	Rus	Fr	Ger	Jap	NKor	SKor	SArabia	Isr	India	UK	Pak
US		1	42		6	1	1	8	1	1	9	6	6
China	15					6					9		
Russia	42			1	2	1							
France	8		13									2	
Germany	6		2							4			
Japan	7	1					3	1					
N. Korea	3												
S. Korea	5					1	5						
Saudi Arabia													
Israel				43							1	1	
India	6	3										1	47
UK	8		1								1		
Pakistan	14										174		

Diagonals are empty. ICEWS Dataverse https://dataverse.harvard.edu/dataverse/icews, extracted by author

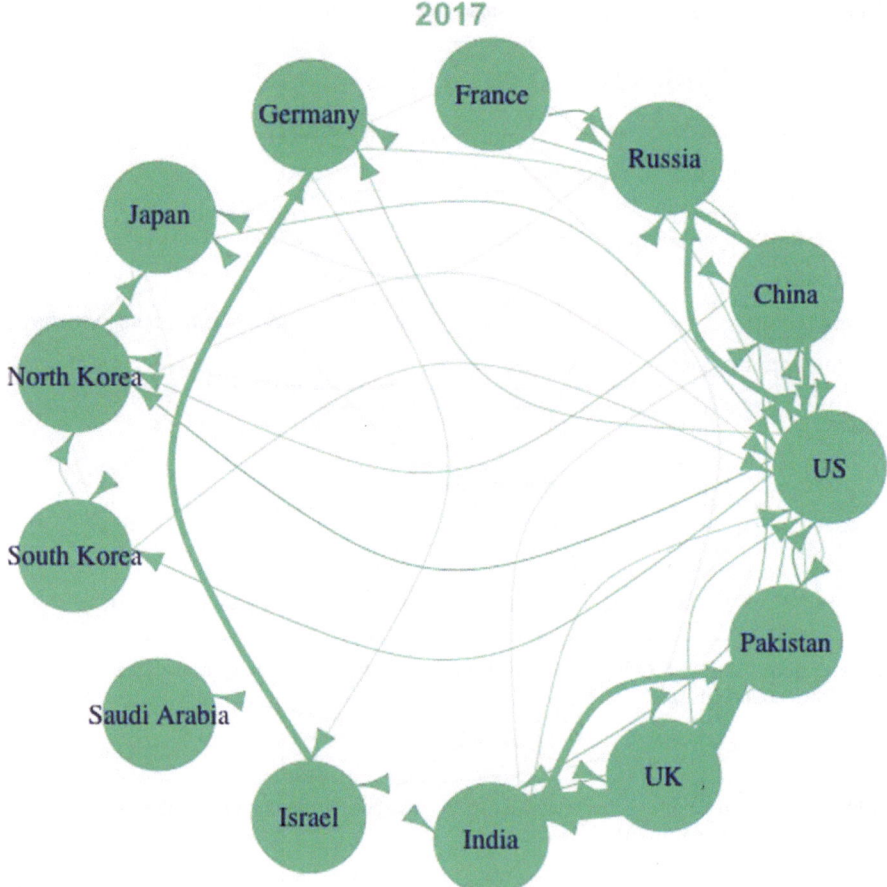

Fig. 6.4 Richardson's $\kappa_{i,j}$ visualized as a weighted network using ICEWS material conflict data for 2017

In these data, Pakistan receives the most material conflict, followed by the US, India, Russia, and China. In terms of sending material conflict, India, followed by the US, far outpaced any of the other countries, though Russia and Pakistan also have sizable material conflict patterns. I display these data in Fig. 6.4 as a network, similar to Fig. 6.1.

This shows reactivity among India, Pakistan, and the UK that is quite strong, as well as among the US, China, Russia clustering. Also, there is quite a bit of conflict from Israel to Germany. All the other reactivities are at a low level.

In order to make this representation compatible with the Richardson $\kappa_{i,j}$ matrix, two things need to be accomplished. First among them is to provide diagonal elements. Then, the matrix needs to be appropriately scaled. Turning to the first task we have already established that the growth rate of military spending varies between 9% for Saudi Arabia to about zero percent for the lowest countries.

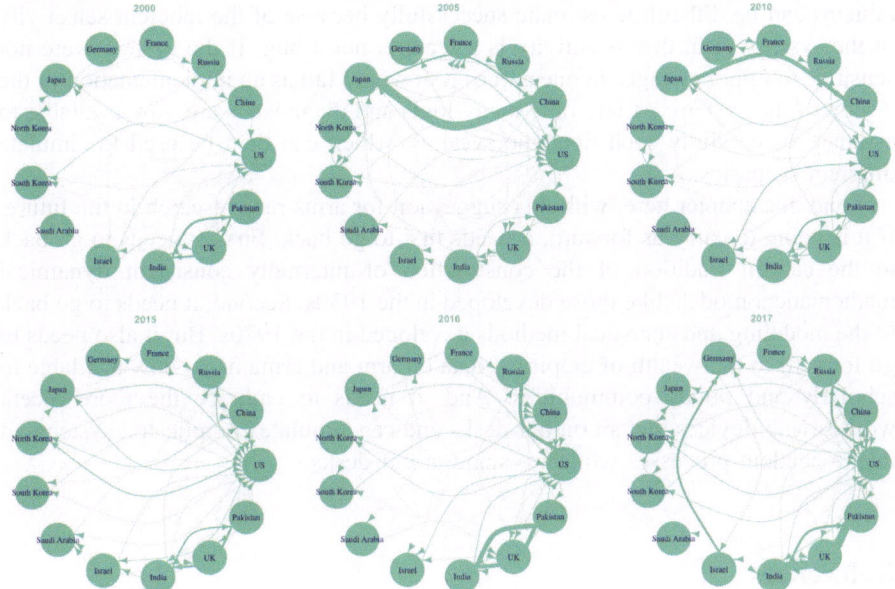

Fig. 6.5 Different annual $\kappa_{i,j}$ representation

This suggest that the diagonals need to off-set, more or less, the sum of the row reactivities. But for now, I set aside this task, for which I suggest a solution below. I focus herein on the year 2017, but it should be obvious that these will fluctuate from year to year, as shown in Fig. 6.5 which has representations for various years from 2000 to 2017.

It is straightforward to implement this 13-nation arms race for 2017 using either Python or R computer libraries that do all the heavy lifting.[7] Because we have access to the SIPRI data for all these countries (more or less), we have enough information to estimate the parameters of a differential or difference equation system, with embedded dependencies. There are two approaches to this, one is to use network methods for the difference equation system (Hoff, 2015; Hoff et al., 2015). The second is to simply use the simulation itself to generate potential parameter values, driving the simulation successively closer to the real data trajectories. This later approach was developed in the 1970s (see, for example, Allan, 1983, or for a Bayesian version, Raftery & Zeh, 1993) and is described more recently in Cellier & Greifeneder (2013).

It is straightforward to use the R packages deSolve and FME to simulate and estimate the parameters for dynamical systems of differential equations.[8] Such

---

[7]Email me, if you would like a copy, to be available in the fullness of time on my github.

[8]Many tools also exist for this in other modeling/statistical framework, e.g., MATLAB among many others.

systems can be difficult to estimate successfully because of the inherent sensitivity of the systems. But this sensitivity is a feature, not a bug. If the system were not sensitive to minor changes in one part of it, it would fail as an implementation of the design of the system. At any rate, many tools and diagnostics are now available to estimate successfully such dynamic systems, which can then be used to simulate different scenarios.

I end this chapter here, with this suggestion for arms race research in the future. If it is going to carry us forward, it needs first to go back. First, it needs to go back to the careful tradition of the construction of internally consistent dynamical mathematical models like those developed in the 1930s. Second, it needs to go back to the modeling and statistical methods developed in the 1970s. But it also needs to go forward to the wealth of empirical data on arm and armaments now available to scholarly and policy communities. And, it needs to embrace the more recent work-saving devices that sit on our desks and can simulate complicated systems of interdependent processes while we stand at our desks.

References

Allan, Pierre (1983) *Crisis Bargaining and The Arms Race: A Theoretical Model.* Cambridge, MA: Ballinger.

Boschee, Elizabeth; Jennifer Lautenschlager, Sean O'Brien, Steve Shellman, James Starz & Michael D Ward (2015) ICEWS coded event data. Harvard Dataverse Network, http://dx.doi.org/10.7910/DVN/28075.

Caspary, William R (1967) Richardson's model of arms races: Description, critique, and an alternative model. *International Studies Quarterly* 11(1): 63–88.

Cellier, Francois E & Jurgen Greifeneder (2013) *Continuous System Modeling.* Springer.

Choucri, Nazli & Robert C North (1972) Dynamics of international conflict. *World Politics* 24(2): 80–122.

Diehl, Paul F (2020) What Richardson got right (and wrong) about arms races and war. Ch. 4 in this volume.

Gleditsch, Nils Petter & Olav Njølstad (eds) (1990) *Arms Races: Technological and Political Dynamics.* London: Sage.

Gleditsch, Kristian Skrede & Nils B Weidmann (2020) From hand-counting to GIS: Richardson in the information age, Ch. 7 in this volume.

Hess, George D (1995) An introduction to Lewis Fry Richardson and his mathematical theory of war and peace. *Conflict Management and Peace Science* 14(1): 77–113.

Hoff, Peter D (2015) Multilinear tensor regression for longitudinal relational data. *Annals of Applied Statistics* 9(3): 1169–1193.

Hoff, Peter D; Bailey Fosdick, Alex Volfovsky & Yanjun He (2015) Amen: Additive and multiplicative effects models for networks and relational data. R package version 1.1. https://github.com/pdhoff/amen.

Lambelet, Jean-Christian; Urs Luterbacher & Pierre Allan (1979) Dynamics of arms races: Mutual stimulation vs self-stimulation. *Conflict Management and Peace Science* 4(1): 49–66.

Luterbacher, Urs (1974) *Dimensions historiques de modèles dynamiques de conflit: application aux processus de course aux armements, 1900–1965,* 2. Geneva: Institut universitaire de hautes études internationales.

McClelland, Charles A (1961) The acute international crisis. *World Politics* 14(1): 182–204.

McClelland, Charles A & Gary Hoggard (1969) Conflict patterns in the interactions among nations. In: James N Rosenau (ed) *International Politics and Foreign Policy*. New York: Free Press, 711–724.

NOAA National Weather Service (2019) Aviation Weather Center, https://www.aviationweather.gov/.

North, Robert C; Ole R Holsti, M George Zaninovich & Dina A Zinnes (1963) *Content Analysis: A Handbook with Applications for the Study of International Crisis*. Evanston, IL: Northwestern University Press.

O'Brien, Sean P (2010) Crisis early warning and decision support: Contemporary approaches and thoughts on future research. *International Studies Review* 12(1): 87–104.

OEDA (2016) Open event data alliance: Phoenix data base. https://github.com/openeventdata.

Pipes, Richard (1986) Team B: The reality behind the myth. *Commentary* October, www.commentarymagazine.com/articles/team-b-the-reality-behind-the-myth/.

Raftery, Adrian E & Judith E Zeh (1993) Estimation of bowhead whale, balaena mysticetus, population size (with discussion). In: Constantine Gatsonis, James S Hodges, Robert E Kass & Nozer D Singpurwalla (eds) *Case Studies in Bayesian Statistics*. New York: Springer, 163–240.

Richardson, Lewis F (1922) *Weather Prediction by Numerical Process*. Cambridge, MA: Cambridge University Press.

Richardson, Lewis F (1960a) *Arms and Insecurity*. Pittsburgh, PA: Boxwood.

Richardson, Lewis F (1960b) *Statistics of Deadly Quarrels*. Pittsburgh, PA: Boxwood.

Schrodt, Philip A (1981) *Preserving Arms Distributions in a Multi-Polar World*. Monograph Series in World Affairs, Graduate School of International Studies 18(4). Denver, CO: University of Denver Press.

Schrodt, Philip A (2015) Event data in forecasting models: Where does it come from, what can it do? Unpublished manuscript.

Schrodt, Philip A, Deborah J Gerner & Ömür Yilmaz (2009) Conflict and mediation event observations (CAMEO): An event data framework for a post-Cold War world. In: Jacob Bercovitch & Scott S Gartner (eds) *International Conflict Mediation: New Approaches and Findings*. New York: Routledge, 287–304.

SIPRI Yearbook 2018: Armaments, Disarmament and International Security. Oxford: Oxford University Press.

Smith, Ron P (2020) The influence of the Richardson arms race model. Ch. 3 in this volume.

Wallace, Michael D (1979) Arms races and escalation: Some new evidence. *Journal of Conflict Resolution* 23(1): 3–16.

Ward, Michael D (1984) Differential paths to parity: A study of the contemporary arms race. *American Political Science Review* 78(2) 297–317.

Ward, Michael D (1985) Simulating the arms race. *Byte* 256(October): 213–222.

Weart, Spencer R (2008) *The Discovery of Global Warming*. 2nd ed. Cambridge, MA: Harvard University Press.

Weart, Spencer R (2019) General circulation models of climate, https://history.aip.org/climate/pdf/Gcm.pdf.

Wilkinson, David (1980) *Deadly Quarrels: Lewis F Richardson and the Statistical Study of War*. Berkeley, CA: University of California Press.

Michael D. Ward, b. 1948, Ph.D. (Northwestern University, 1977); Emeritus Professor, Duke University; Affiliate Professor University of Washington; most recent book (with John Ahlquist): *Maximum Likelihood for Social Science* (Cambridge University Press, 2019). Founder and CEO, Predictive Heuristics, michael.don.ward@gmail.com.

Chapter 7
From Hand-Counting to GIS: Richardson in the Information Age

Kristian Skrede Gleditsch and Nils B. Weidmann

Abstract Richardson made pioneering contributions to the study of geography and its influence on social and political dynamics. We use the research of Richardson as a point of departure to examine how Geographic Information Systems (GIS) technology and spatial data provide opportunities to answer old and new questions in conflict research. There is an enduring interest in how geographical features influence political interactions and outcomes and increasing attention to how key factors highlighted vary spatially both within countries and beyond national boundaries. We focus on key motivations for using spatially disaggregated data and show how such data can help advance core research questions, drawing on examples from the study of violent conflict.

7.1 Introduction

Lewis Fry Richardson collected an influential early database on 'deadly quarrels' and made prominent contributions to modelling interactions such as arms races using differential equations (Hess, 1995; Nicholson, 1999; Richardson, 1960a). Less well known is Richardson's pioneering work on geography, examining topics such as the relationship between borders and conflict, developing measures of territorial properties such as 'compactness', as well as a number of interesting observations on the political implications and origins of borders (Richardson, 1960b, 1961). For example, Richardson noted how administratively determined internal borders tended to look very different from 'natural' external borders. While

---

The original version of this chapter was revised: The name "Gregory D. Hess" has been corrected to "George D. Hess" in Reference list. The correction to this chapter is available at https://doi.org/10.1007/978-3-030-31589-4_12

This chapter is a shortened version of Gleditsch & Weidmann (2012), with some revision and updating. Reprinted with permission.

the latter tend to follow physical features such as rivers or mountain ranges, the former often take the form of straight lines clearly drawn directly on a map, usually without regard for natural features (see also Mandelbrot, 1967 on the scale effect of borders noted by Richardson). Richardson further noted that there were no instances of four independent states meeting in a single point, such as the Four Corners area of the United States. He attributed this to the role of warfare in shaping borders and the difficulty of maintaining border arrangements that would be difficult to defend militarily. The Caprivi Strip, a protruding part of northeastern Namibia, is sometimes held up as a contemporary counterexample to Richardson's observation. Namibia and Zimbabwe do not appear to be contiguous, even if both border the Zambezi river. Still, this strangely shaped area emerged from complex treaties between the UK and Germany, and has seen considerable conflict and contention, consistent with Richardson's core intuition.

Richardson had limited tools at his disposal when writing in the 1930s, and most of his geographical computations were done by hand. In this chapter, we review research picking up the gavel from Richardson, using the tools of the information age and advances in Geographic Information Systems (GIS). We show that spatial data can provide important new insights in conflict research, enhance theory-measure correspondence, and inform models of spatial variation and processes.

7.2   The GIS Revolution in the Social Sciences

Contemporary research often uses GIS to examine smaller and more fine-grained data. 'Spatial disaggregation' is often employed to move below the country-level and use local indicators to better approximate the specific actors and mechanisms of interest (Cederman & Gleditsch, 2009). For example, researchers have focused on the conflict zones in civil wars and examining local correlates of violence. Moreover, GIS can help provide access to new information relevant to conflict processes such as the spatial concentration of ethnic groups or proximity to conflict. In addition, spatial datasets can be used to capture phenomena that are plausibly exogenous to social processes such as for example weather or terrain, which can be extremely useful for causal identification.

The term 'Geographic Information System' denotes a family of software tools that allow for the collection, visualization and analysis of spatial data. GIS analysis extends beyond creating maps, and a key promise lies in the ability to compute spatial indicators. Some computations operate on a single dataset (or 'layer') as input. For example, only one input layer of country borders is required to compute minimum distances between countries. More complex operations use the spatial co-occurrence of information contained in different datasets. For example, we can compute an indicator of terrain ruggedness by overlaying data on units with information about territorial elevation, and then examine it is relationship with conflict events.

Richardson's (1960b) dataset on 'deadly quarrels' contained much information but did not provide very previse spatial information. There has been a rapid growth in GIS use over the last decade, and many GIS datasets cover issues relevant to conflict researchers, with explicit spatial information. Furthermore, it is straight-forward to collect new GIS data for spatial analysis. Spatial data can be represented either in a vector or a raster format. Vector formats are typically used for discrete spatial entities, while rasters represent a continuous variable over space. Our discussion here must be selective, and we refer to Gleditsch & Weidmann (2012) and Ward & Gleditsch (2018) for more detailed general overviews.

We focus on a worked example of measuring horizontal inequality across ethnic groups within countries, based on Cederman et al. (2011). There is a long research tradition on whether grievances generated by economic inequality increase conflict (Gurr, 1970). Earlier research found more political protest under higher inequality (Muller & Seligson, 1987), but many studies of civil war find no clear relationship between measures of interpersonal income inequality and conflict (Collier & Hoeffler, 2004). However, 'vertical' inequality between individuals is conceptually distinct from 'horizontal' inequalities that coincide with other salient cleavages such as ethnic divisions. Many argue that the latter is more likely to spur violent mobilization, given the important relationship between ethnic groups and opportunities for collective action (see Cederman et al., 2011; Stewart, 2008; Østby, 2008).

The first building block in our example is data on national boundaries. Our CShapes dataset provides historical country borders as vector polygons for the post-World War II period (Weidmann et al., 2010). Even if borders are not of primary interest, these data allow linking other variables of interest to spatial referents and create maps or spatial measures. In addition, the associated CShapes R package allows the user to compute derived measures from the country polygons, such as the minimum distance between countries (Weidmann & Gleditsch, 2010).

GeoEPR (Wucherpfennig et al., 2011) is a spatial extension to the Ethnic Power Relations dataset (Wimmer et al., 2009), and provides a dynamic spatial coding of ethnic group settlement regions, with polygon 'lifespans'. GeoEPR makes it possible to rely on GIS techniques (in particular, overlays) to derive a variety of spatial and non-spatial indicators for ethnic groups.

The G-Econ data provides estimate of sub-national economic activity for 1-degree grid cells (Nordhaus, 2006). We can derive per capita and inequality measures for ethnic groups by overlying the G-Econ data with the GeoEPR settlement polygons and demographic information from the Gridded Population of the World,[1] a raster with a resolution of 2.5 arc-minutes.

Figure 7.1 illustrates the construction of the measures for Yugoslavia. Overlaying the G-Econ data and the GeoEPR data gives us a measure of total economic activity by group settlement area. We can then consider group inequality by comparing per-capita wealth for each group with the national average, with values above 1 for relatively more affluent groups and values below 1 for poorer groups (Fig. 7.1, right). The ratios indicate that Albanians in Kosovo are on average poorer than the national average, while the Croats and the Slovenes are wealthier.

---

[1]http://sedac.ciesin.columbia.edu/data/collection/gpw-v4.

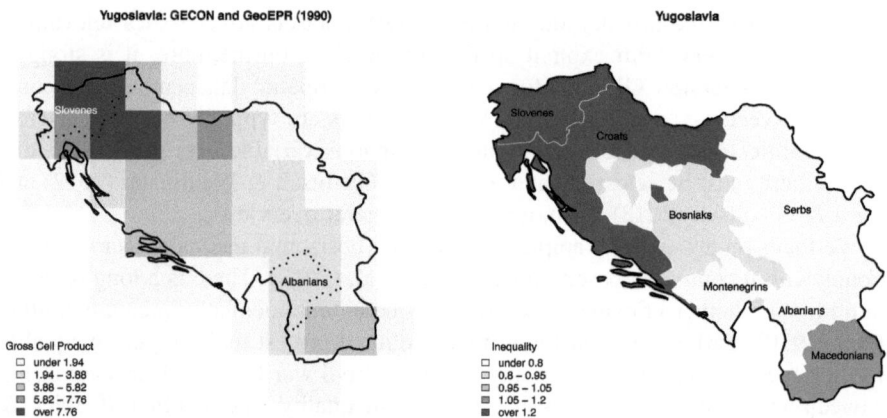

Fig. 7.1 Example for the computation of the group wealth indicator. The G-Econ dataset on economic performance is overlaid with the group settlement regions from Geo-EPR (left). Aggregating the (partial) G-Econ cell values by group results in wealth estimates at the group level (shown as proportions of the national average, right). *Source* The authors

Cederman et al. (2011) use these ratios in a global analysis and find that large inequalities along ethnic lines are associated with a higher risk of ethnic conflict, both for relatively disadvantaged and privileged groups.

Cederman et al. (2011) use information reflecting whether actors in armed conflict as described by Gleditsch et al. (2002) are linked to ethnic groups, but the conflict data are not actually spatial. We now have a host of spatial datasets on political violence. In an early attempt to spatially reference civil war, Buhaug & Gates (2002) coded a 'conflict zone' based on the smallest circle surrounding all violent events in a given country and year and examined how location and scope varied by geographical characteristics and country attributes (see also Hallberg, 2012). Other data sources attempt to provide precise information on the individual events that make up a 'war' or conflict episode, tagged each event with temporal and spatial coordinates in a point vector representation. The Armed Conflict Location and Event Dataset ACLED (Raleigh et al., 2010, www.acleddata.com) as well as the Georeferenced Event Data (GED) of the Uppsala Conflict Data Project (Sundberg & Melander, 2013) provide incident level data for civil wars and non-state actor conflict. For interstate conflict, the MIDLOC dataset (Braithwaite, 2010) reports the onset of each episode in the Militarized Interstate Disputes dataset. The Social Conflict Analysis Database (SCAD, Salehyan et al., 2012) provides spatial coordinates for non-violent and less organized violent events, including riots, strikes, and protests.

It is increasingly popular to use GIS datasets for information on geographic or environmental characteristics that can help support casual inference. Concern over the potential endogeneity of various economic and political explanatory factors have led researchers to look to geographic or environmental characteristics as potentially exogenous sources of variation or 'deep' determinants. For example,

Miguel et al. (2004) use rainfall data as plausible exogenous shocks in economies dominated by rainfed agriculture to improve causal identification of the effects of economic growth on conflict. Michalopoulos (2012) argues that soil quality and elevation provide exogenous sources of ethnic diversity, as higher regional variation should reduce migration and lead to a higher number of ethnic groups. Many environmental variables can be measured using satellite imagery. The GTOPO30 dataset is a global raster data on territorial elevation, measured at the level of grid cells with a resolution of 30 arc-seconds.[2] Estimates of rainfall and related variables are provided in raster formats by the Global Precipitation Climatology Project, available at https://rda.ucar.edu/datasets/ds728.3/.

Disaggregated spatial data come in different resolutions and combining different data sources usually require scaling to some common resolution. The PRIO-GRID project (Tollefsen et al., 2012) provides a standardized grid structure that integrates different data sources to a common set of geographical cells with a resolution of 0.5 decimal degrees (roughly 50 km at the equator). Buhaug et al. (2011) provide a grid cell analysis of local economic characteristics and the initial onset event in a conflict, controlling for a host of social and political factors believed to influence the risk of conflict.

Rather than relying on existing spatial datasets, researchers may create new spatial datasets by either recording spatial coordinates when data are collected, or appending spatial information to existing, non-spatial data, a step that is usually called 'geo-referencing'. Spatial coordinates for observations can be derived by the Global Positioning System (GPS), where a GPS receiver with the help of satellites can determine geographic position with a high level of accuracy. Surveys often record the geographic position of a respondent or an interview, which allows linking these to other GIS layers. The Demographic and Health Surveys project, for example, conducting surveys on various living standards and health related outcomes for households, routinely attaches GPS coordinates (www.measuredhs.com). For example, Hegre et al. (2009) use DHS data to approximate geographical variation in poverty by grids in Liberia.

Existing, non-spatial datasets can be made GIS-compatible in different ways. Event datasets from news reports usually obtain spatial coordinates by converting place names into geographic coordinates. The location of the village or city mentioned in media reports can be found with the help of gazetteers, a list of place names and their spatial coordinates. Useful gazetteers include the Falling Rain database (www.fallingrain.com/world/index.html), or the GEOnet Names Server at the NGA (http://geonames.nga.mil/gns/html/). Since the spelling of place names often is not standardized, the JRC Fuzzy Gazetteer (http://isodp.hofuniversity.de/fuzzyg/query/) is particularly convenient, since it retrieves place names even if the spelling does not match perfectly.

Alternatively, GIS databases can be created by converting existing maps into GIS-compatible formats. After scanning maps and aligning the map correctly with

---

[2]https://eros.usgs.gov/#/Find_Data/Products%20and_Data_Available/gtopo30_info.

the spatial reference system used by the GIS, the spatial features of interest can be extracted manually or by applying a feature recognition algorithm (see Longley et al., 2010). Vanzo (1999), for example, geocodes historical maps reflecting boundary changes to examine to what extent post-conflict borders reflect a tendency towards greater territorial compactness.

7.3   GIS and Spatial Data Analysis

Our first example of GIS in analysis considers how information on the location of violence can help inform research on the causes and consequences of conflict. Much research has considered 'civil war' as a dichotomous outcome, where states are either 'at war' or not over some specific period. However, civil wars rarely engulf entire countries and come in many different degrees, both in terms of the severity and geographical scope of fighting. The *Conflict Sites* dataset expands the Uppsala Armed Conflict Data to a geographical representation of the zone where violence takes place, using a polygon representation. Figure 7.2 displays two important examples of variation in the distribution in civil wars. For example, the Chechen War in 1995 (left panel) is confined to a relatively small and peripheral part of the territory of Russia, and clearly does not extend to the entire country. The conflict zone is small and unlikely to influence national figures, and national level data are unlikely to reflect the local impact of the conflict in the region. Conversely, the Democratic Republic of Congo in 2007 (right panel) experiences two distinct civil wars that take place in completely different parts of the country, with the National Congress for the Defence of the People (CNDP, a Tutsi dominated organization) in the East, and the Bundu Dia Kongo, which claims to represent the Kongo people, in the West. Treating the country at large as 'at war' distracts our attention from the distinct actors and conflictual interactions taking place.

Fig. 7.2 Conflict polygons for the Chechen War in Russia in 1995 (left), and the 2007 civil wars in the Democratic Republic of Congo (right). *Source* The authors

GIS provide opportunities for more systematic analyses of how spatially varying features influence conflict. If politics are local, then the causes of the conflict are more likely to reflected in the characteristics of the areas where they occur rather than features of state at large. By construction, many conventional country-level measures such as Gross Domestic Product per capita or ethnic fractionalization are averages that reflect population density and will not reflect variation within countries. Buhaug & Lujala (2005) compare conflict zones to other areas within the same country without conflict using GIS data, and demonstrate that conflict zones tend to be very different from other areas of a country. Since civil wars tend to be fought by small groups, often in thinly populated peripheral areas, the risk of conflict may be better reflected by 'worst case' indicators, or measures of the geographical areas most likely to see conflict, rather than population weighted measures (Buhaug et al., 2014). More generally, researchers should think carefully about correspondence between actual measures and the underlying theoretical concepts. Just because a particular measure is available or is used in existing research it does not necessarily follow that it is a suitable indicator for testing a particular argument.

Buhaug & Gates (2002) examine a number of hypotheses on possible factors that may account for variation in the size or geographical scope of conflict zones. They find strong evidence that the presence of natural resources within conflicts and their overall duration influence the geographical scope of conflicts, and their results suggest a possible endogenous relationship between the peripheral location of conflict and its geographical scope.

The consequences of conflict are likely to be proportional to their magnitude. Although civil wars can be shown to have a negative impact on social and economic development, it seems unreasonable to expect that conflicts with a limited geographical scope would have identical consequences to large conflicts with broad geographical reach. As a supplement to national level studies (e.g., Bozzoli et al., 2010), many studies that look at the impact of conflict at the household or individual level using survey data (Verwimp et al., 2009).

Beyond static features, spatial data can also be used to analyze the dynamics of change over time, such as the diffusion of conflict. Many important mechanisms can create spatial dependence between actors or locations and increase the risk of conflict. More generally, if ongoing conflict in one country can affect the risk of conflict in other states, the individual conflict outbreaks are not independent as the outcomes are shaped by events and outcomes in other, connected observations.

Much of the research on civil war has adopted a 'closed polity' approach, assuming that the relevant causes of internal conflicts must be found within the boundaries of the country experiencing conflict (Gleditsch, 2007). However, there are strong theoretical reasons why the risk of conflict may be shaped by events and features in other states, especially neighboring countries. For example, many conflicts involve demands for autonomy or independence by ethnic communities, who often reside in multiple countries (Cederman et al., 2009; Lake & Rothchild, 1998). The decision to contest the state militarily can be influenced by experiences of the group in another state, or the ability to rely on financial or military support from kin in another state. An ongoing civil war in a neighboring country can increase the availability of arms and recruits and make it relatively less costly to mobilize insurgencies (Lischer, 2005; Salehyan & Gleditsch, 2006). Hostile relations between states can give governments incentives to support insurgencies in a neighboring state to undermine their rival (Davis & Moore, 1997; Salehyan et al., 2011).

GIS data can used to examine whether conflict affected areas cluster geographically and how they evolve over time. The Great Lakes Region of Africa in the 1990s is often cited as an example of a cluster of interdependent civil wars (Prunier, 2008; McNulty, 1999). Figure 7.3 displays the Conflict Sites polygons for the years 1991 to 1999. The civil war in Rwanda erupted when the Rwandan Patriotic Front (RPF) invaded from Uganda in 1991, where a Tutsi refugee population of about 200,000 individuals had organized militarily, with assistance from Ugandan authorities and partial integration of rebel forces in the regular army.

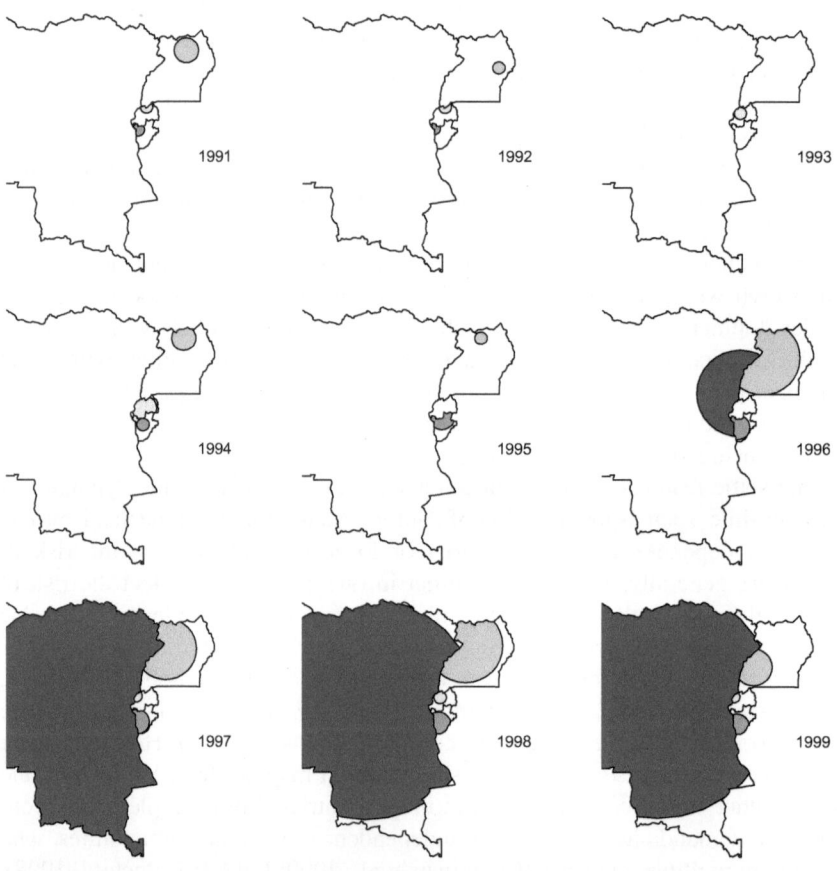

Fig. 7.3 Conflict polygons in the Great Lakes Region of Africa, 1991–99. *Source* The authors

The conflict polygon in Rwanda is clearly located on the border, reflecting the important ties to Uganda, as the RPF on occasion retreated into Uganda to regroup and rearm during the initial period.

The maps for the subsequent years reflect the escalation to encompass the whole country around the Rwandan genocide in 1994 and the eventual RPF victory. The Rwandan civil war generated a major Hutu refugee crisis in neighboring countries, in particular Zaire, where the refugee camps provided a fertile environment for a Hutu insurgent movement, the Rassemblement Démocratique pour le Rwanda. In response, the Kagame government in Rwanda supported an insurgent group in Zaire, the Alliance of Democratic Forces for the Liberation of Congo (AFDL) led by Kabila. This is reflected in the conflict polygon in Eastern Zaire in 1996, again clearly on the border with Rwanda. The 1997 maps show the subsequent escalation, where the AFDL overthrows Mobutu in 1997 and declares a new Democratic Republic of the Congo (DRC). The maps for the subsequent years demonstrate how peace has remained elusive in the region, and possibly may have fueled an escalation of the civil war in Uganda.

These spatial representations are in line with Richardson's emphasis on borders as an opportunity for interaction and the diffusion of conflict (Richardson, 1960b, 1961; see also Siverson & Starr, 1991). Although this is just a single case, many global studies find considerable support for the importance of spatial proximity in conflict diffusion (Bosker & de Ree, 2014; Gleditsch, 2007; Ward & Gleditsch, 2002). Hegre & Sambanis (2006) report neighboring conflict as one of the key features with a robust positive influence on the risk of civil war in their sensitivity analysis. Other researchers have estimated the effects of specific mechanisms or transnational linkages, including transborder ethnic kin (Bosker & de Ree, 2014; Cederman et al., 2009), or refugee flows (Salehyan & Gleditsch, 2006). More recent studies have looked at diffusion within individual conflicts. Schutte & Weidmann (2011) distinguish between two types of diffusion, relocation and escalation, each of which is the result of a particular type of warfare. They find that civil wars primarily exhibit escalation diffusion as a result of irregular warfare without conventional front lines. Weidmann & Ward (2010) consider the spatial and temporal diffusion for conflict events in Bosnia, demonstrating that violence is likely to recur over time and spread spatially, and showing that taking advantage of this information can substantial improve the ability to forecast conflict.

7.4   Conclusion

We started by arguing that GIS and the increasing availability of spatial data provide many opportunities for advancing research on the spatial features of conflict and political interactions highlighted in the pioneering research by Richardson. There is an interesting analogy here to Richardson's (1922) work on weather forecasting, which proposed a system based on solving differential equations that was not computationally feasible at the time. Subsequent advances in computing, however, have vindicated Richardson's ideas (Lynch, 2006). The first modern computer ENIAC generated a weather forecast in 1950, and similar models are today used extensively for weather forecasting and modeling climate change. Of course, meteorological processes are different from social interactions that often involve strategic behavior and responses, but spatial variation and features can be incorporated in many strategic formal models (e.g., Fujita et al., 2001; Alesina & Spolaore, 2003) or computational models (e.g., Turchin, 2003; Epstein, 2007).

Our overview provides strong support for the claim that GIS and spatial data has helped advance research on spatial features and political interactions and outcomes in the spirit of Richardson's initial efforts. Spatial data have helped facilitate new approaches to the study of inequality and conflict, which at least to some provide a vindication of the role of grievances in civil war often dismissed in other research. We have learned important things about the risks of conflict diffusion, as well as of specific conditions where conflicts are more or less likely to generate instability in other countries. Although much work remains to be done, we believe that Richardson would have been very pleased to see the results of existing research using GIS and spatial data.

The spatial perspective is not just a question of tools and techniques, but also helps foster a substantively novel theoretical approach for understanding political events and outcomes. Whereas much research often takes states as predetermined units with fixed boundaries, Richardson's work on borders alerts us to the endogenous nature of borders, and how present-day borders reflect historical and political processes that have generated and preserved borders (Alesina et al., 2011; Englebert et al., 2002). Whereas much comparative research traditionally may have treated individual countries as independent units, interdependence is an essential characteristic of a globalizing world.

References

Alesina, Alberto & Enrico Spolaore (2003) *The Size of Nations*. Cambridge, MA: MIT Press.

Alesina, Alberto; William Easterly & Janina Matuszeski (2011) Artificial states. *Journal of the European Economic Association* 9(2): 246–277.

Bosker, Maarten & Joppe de Ree (2014) Ethnicity and the spread of civil war. *Journal of Development Economics* 108: 206–221.

Bozzoli, Carlos; Tilman Brück & Simon Sottsas (2010) A survey of the global economic costs of conflict. *Defence and Peace Economics* 21(2): 165–176.

Braithwaite, Alex (2010) MIDLOC: Introducing the Militarized Interstate Dispute Location dataset. *Journal of Peace Research* 47(1): 91–98.

Buhaug, Halvard & Scott Gates (2002) The geography of civil war. *Journal of Peace Research* 39 (4): 417–433.

Buhaug, Halvard & Päivi Lujala (2005) Accounting for scale: Measuring geography in quantitative studies of civil war. *Political Geography* 24(4): 399–418.

Buhaug, Halvard; Kristian Skrede Gleditsch, Helge Holtermann, Gudrun Østby & Andreas Forø Tollefsen (2011) It's the local economy, stupid! Geographic wealth dispersion and conflict outbreak location. *Journal of Conflict Resolution* 55(5): 814–884.

Buhaug, Halvard; Lars-Erik Cederman & Kristian Skrede Gleditsch (2014) Square pegs in round holes: Grievances, inequalities, and civil war. *International Studies Quarterly* 58(2): 418–431.

Cederman, Lars Erik & Kristian Skrede Gleditsch (2009) Special Issue on 'Disaggregating civil war'. *Journal of Conflict Resolution* 53(4): 487–495.

Cederman, Lars-Erik; Luc Girardin & Kristian Skrede Gleditsch (2009) Ethno-nationalist triads: Assessing the influence of kin groups on civil wars. *World Politics* 61(3): 403–437.

Cederman, Lars-Erik; Nils B Weidmann & Kristian Skrede Gleditsch (2011) Horizontal inequalities and ethno-nationalist civil war: A global comparison. *American Political Science Review* 105(2): 478–495.

Collier, Paul & Anke Hoeffler (2004) Greed and grievance in civil war. *Oxford Economic Papers* 56: 663–595.

Davis, David R & Will H Moore (1997) Ethnicity matters: Transnational ethnic alliances and foreign policy behavior. *International Studies Quarterly* 41(1): 171–184.

Englebert, Pierre; Stacey Tarango & Matthew Carter (2002) Dismemberment and suffocation: A contribution to the debate on African boundaries. *Comparative Political Studies* 35(10): 1093–1118.

Epstein, Joshua M (ed) (2007) *Generative Social Science: Studies in Agent-Based Computational Modeling*. Princeton, NJ: Princeton University Press.

Fujita, Masahisa; Paul Krugman & Anthony J Venables (2001) *The Spatial Economy: Cities, Regions, and International Trade*. Cambridge, MA: MIT press.

Gleditsch, Kristian Skrede (2007) Transnational dimensions of civil war. *Journal of Peace Research* 44(3): 293–309.

Gleditsch, Kristian Skrede & Nils B Weidmann (2012) Richardson in the information age: Geographic information systems and spatial data in international studies. *Annual Review of Political Science* 15: 461–481. Revised version in German: Geodaten und deren Analyse in der Politikwissenschaft. In: Claudius Wagemann, Achim Goerres & Markus Siewert (eds) (2018) *Handbuch Methoden der Politikwissenschaft*. Frankfurt: Springer, https://doi.org/10.1007/978-3-658-16937-4_23-1.

Gleditsch, Nils Petter; Peter Wallensteen, Mikael Eriksson, Margareta Sollenberg & Håvard Strand (2002) Armed conflict 1946–2001: A new dataset. *Journal of Peace Research* 39(5): 615–637.

Gurr, Ted R (1970) *Why Men Rebel*. Princeton, NJ: Princeton University Press.

Hallberg, Johan Dittrich (2012) PRIO Conflict Sites 1989–2008: A geo-referenced dataset on armed conflict. *Conflict Management and Peace Science* 29(2): 219–32.

Hegre, Håvard & Nicholas Sambanis (2006) A sensitivity analysis of the empirical literature on civil war onset. *Journal of Conflict Resolution* 50(4): 508–535.

Hegre, Håvard; Gudrun Østby & Clionadh Raleigh (2009) Poverty and civil war events: A disaggregated study of Liberia. *Journal of Conflict Resolution* 53(4): 598–623.

Hess, George D (1995) An introduction to Lewis Fry Richardson and his mathematical theory of war & peace. *Conflict Management and Peace Science* 14(1): 77–113.

Lake, David A & Donald Rothchild (eds) (1998) *The International Spread of Ethnic Conflict: Fear, Diffusion, and Escalation*. Princeton, NJ: Princeton University Press.

Lischer, Sarah Kenyon (2005) *Dangerous Sanctuaries: Refugee Camps, Civil War, and the Dilemmas of Humanitarian Aid*. Cornell, NY: Cornell University Press.

Longley, Paul A; Michael F Goodchild, David J Maguire & David W Rhind (2010) *Geographic Information Systems and Science*. Chichester: Wiley.

Lynch, Peter (2006) *The Emergence of Numerical Weather Prediction: Richardson's Dream*. Cambridge: Cambridge University Press.

Mandelbrot, Benoit (1967) How long is the coast of Britain? Statistical self-similarity and fractional dimension. *Science* 156(3775): 636–638.

McNulty, Mel (1999) The collapse of Zaire: Implosion, revolution, or external sabotage? *Journal of Modern African Studies* 39(1): 53–82.

Michalopoulos, Stelios (2012) The origins of ethnolinguistic diversity. *American Economic Review* 102(4): 1508–1039.

Miguel, Edward; Shanker Satyanath & Ernest Sergenti (2004) Economic shocks and civil conflict: An instrumental variables approach. *Journal of Political Economy* 112(4): 725–753.

Muller, Edward N & Mitchell A Seligson (1987) Inequality and insurgency. *American Political Science Review* 87(2): 425–451.

Nicholson, Michael (1999) Lewis Fry Richardson and the study of the causes of war. *British Journal of Political Science* 29(3): 541–563.

Nordhaus, William D (2006) Geography and macroeconomics: New data and new findings. *Proceedings of the National Academy of Sciences USA* 103(10): 3510–3517.

Østby, Gudrun (2008) Polarization, horizontal inequalities and violent civil conflict. *Journal of Peace Research* 45(2): 143–162.

Prunier, Gerard (2008) *Africa's World War: Congo, the Rwandan Genocide, and the Making of a Continental Catastrophe*. Oxford: Oxford University Press.

Raleigh, Clionadh; Andrew Linke, Håvard Hegre & Joakim Karlsen (2010) Introducing ACLED: An armed conflict location and event dataset. *Journal of Peace Research* 47(5): 651–660.

Richardson, Lewis F (1922) *Weather Prediction by Numerical Process*. Cambridge: Cambridge University Press.

Richardson, Lewis F (1960a) *Arms and Insecurity*. Pittsburgh, PA: Boxwood.

Richardson, Lewis F (1960b) *Statistics of Deadly Quarrels*. Pittsburgh, PA: Boxwood.

Richardson, Lewis F (1961) The problem of contiguity: An appendix to Statistics of Deadly Quarrels. *General Systems Yearbook* 6: 140–187. Reprinted in: Oliver M Ashford, H Charnock, PG Drazin, Julian CR Hunt, Paul Smoker & Ian Sutherland (eds) *Collected Papers of Lewis Fry Richardson*, 2 Quantitative Psychology and Studies of Conflict. Cambridge: Cambridge University Press, 577–627.

Salehyan, Idean & Kristian Skrede Gleditsch (2006) Refugee flows and the spread of civil war. *International Organization* 60(2): 335–366.

Salehyan, Idean; Kristian Skrede Gleditsch & David Cunningham (2011) Explaining external support for insurgent groups. *International Organization* 65(4): 709–744.

Salehyan, Idean; Cullen S Hendrix, Jesse Hamner, Christina Case, Christopher Linebarger, Emily Stull & Jennifer Williams (2012) Social conflict in Africa: A new database. *International Interactions* 38(4): 503–511.

Schutte, Sebastian & Nils B Weidmann (2011) Diffusion patterns of violence in civil wars. *Political Geography* 30(3): 143–152.

Siverson, Randolph M & Harvey Starr (1991) *The Diffusion of War: A Study in Opportunity and Willingness*. Ann Arbor, MI: University of Michigan Press.

Stewart, Frances (ed) (2008) *Horizontal Inequalities and Conflict: Understanding Group Violence in Multiethnic Societies*. Houndmills: Palgrave Macmillan.

Sundberg, Ralph & Erik Melander (2013) Introducing the UCDP Georeferenced Event Dataset. *Journal of Peace Research* 50(4): 523–532.

Tollefsen, Andreas Forø; Håvard Strand & Halvard Buhaug (2012) PRIO-GRID: A unified spatial data structure. *Journal of Peace Research* 49(2): 363–374.

Turchin, Peter (2003) *Historical Dynamics: Why States Rise and Fall*. Princeton, NJ: Princeton University Press.

Vanzo, John P (1999) Border configuration and conflict: Geographical compactness as a territorial ambition of states. In: Paul Diehl (ed) *A Road Map to War: Territorial Dimensions of International Conflict*. Nashville, TN: Vanderbilt University Press, 73–112.

Verwimp, Philip; Patricia Justino & Tilman Brück (2009) The analysis of conflict: A micro-level perspective. *Journal of Peace Research* 46(3): 307–314.

Ward, Michael D & Kristian Skrede Gleditsch (2002) Location, location, location: An MCMC approach to modeling the spatial context of war and peace. *Political Analysis* 10(2): 244–260.

Ward, Michael D & Kristian Skrede Gleditsch (2018) *Spatial Regression Models*. Second edition. Thousand Oaks, CA: Sage.

Weidmann, Nils B & Kristian Skrede Gleditsch (2010) Mapping and measuring country shapes: The Cshapes package. *R Journal* 2(1), online at http://journal.r-project.org/archive/2010-1/.

Weidmann, Nils B & Michael D Ward (2010) Predicting conflict in space and time. *Journal of Conflict Resolution* 54(6): 883–901.

Weidmann, Nils B; Doreen Kuse & Kristian Skrede Gleditsch (2010) The geography of the international system: The CShapes dataset. *International Interactions* 36(1): 86–106.

Wimmer, Andreas; Lars-Erik Cederman & Brian Min (2009) Ethnic politics and armed conflict: A configurational analysis of a new global dataset. *American Sociological Review* 74(2): 316–337.

Wucherpfennig, Julian; Nils B Weidmann, Luc Girardin, Lars-Erik Cederman & Andreas Wimmer (2011) Politically relevant ethnic groups across space and time: Introducing the GeoEPR dataset. *Conflict Management and Peace Science* 28(5): 423–437.

Kristian Skrede Gleditsch, b. 1971, Ph.D. in Political Science (University of Colorado, Boulder, 1999); Regius Professor, Department of Government, University of Essex (2005/2018–); Research Associate, Peace Research Institute Oslo (2003–); current research interests: violent and nonviolent conflict, democratization, and political violence, ksg@essex.ac.uk

Nils B Weidmann, b. 1976, Ph.D. in Political Science (ETH Zurich, 2009); Professor of Political Science, University of Konstanz, Germany (2012–) and head of the Communication, Networks and Contention research group; research interests: violent and nonviolent conflict, with a particular focus on the impact of communication and information technology, nils.weidmann@uni-konstanz.de

Chapter 8
Weather, War, and Chaos: Richardson's Encounter with Molecules and Nations

Jürgen Scheffran

Abstract Richardson's pioneering work on modeling conflict and arms races has demonstrated that mathematics can contribute to peace and conflict research, using system dynamics and stability conceptions to study both nature and society. Drawing from limitations and extensions of Richardson's model, including decision rules and chaos in arms races, an integrated modeling framework of social interaction among multiple agents is presented to study conflict phenomena in a complex world. Conditions for instability and chaos are discussed, potentially leading to arms races and violent conflicts, as well as transitions between conflict and cooperation. The model offers a basis for insights into the analysis of potential relationships of natural resources and climate change with social stability and conflict, building bridges between Richardson's research in atmospheric sciences and his work on peace and conflict.

8.1 On Molecules and Nations: Richardson's Scientific Conceptions

Lewis Fry Richardson (1881–1953), a British physicist, psychologist and pacifist, made important contributions to weather forecasting and conflict research and applied approaches and methodologies from physics, mathematics and atmospheric science to social phenomena (Ashford, 1985; Vulpiani, 2014). In particular, the concepts of equilibria and stability which are relevant for differential equations in meteorology and their solutions were transferred to the understanding of arms races (Richardson, 1956: 1247): 'stability is not the same as equilibrium; for on the contrary stable and unstable are adjectives qualifying equilibrium. Thus, an equilibrium is said to be stable, or to have stability, if a small disturbance tends to die away; whereas an equilibrium is said to be unstable, or to have instability, if a small

---

The original version of this chapter was revised: The name "Gregory D. Hess" has been corrected to "George D. Hess" in Reference list. The correction to this chapter is available at https://doi.org/10.1007/978-3-030-31589-4_12

© The Author(s) 2020, corrected publication 2021
N. P. Gleditsch (ed.), *Lewis Fry Richardson: His Intellectual Legacy and Influence in the Social Sciences*, Pioneers in Arts, Humanities, Science, Engineering, Practice 27, https://doi.org/10.1007/978-3-030-31589-4_8

disturbance tends to increase. ... It is the instability which has the disastrous consequences.'

Regarding the notion of small differences versus large impacts, Richardson preempted concepts of chaos and complexity. In a letter to *Nature* he paid attention to the similarities between the behavior of nations and of gas molecules (Richardson, 1946a: 135). Starting from the observation that in a gas 'encounters of two molecules are much more frequent than encounters of three' he explained this by the product of three probability factors relevant in a theory of gas and the political world. He continued: 'Although three factors of the aforesaid sort are likely to appear in the theory of any chaos, yet their particular forms depend on circumstances; so that many varieties of chaos are conceivable. In the political world there were restrictions depending on geography and on sea-power. When they had been formulated, another effect became conspicuous, namely, the infectiousness of local fighting.'

This analysis has been expanded in two separate publications, concerning gases (Richardson, 1946b), and concerning the political world (Richardson, 1946c). In the latter, Richardson tested 13 theories of various degrees of complexity, and derived the interpretive idea of chaos, 'with its characteristic property that complicated events are rarer than simpler events. ... the complicated events are regarded as built up from simpler elements ... The more such elements, the less resultant probability.' (Richardson, 1946c: 138). He highlighted 'chaos, restricted by geography, and further modified by the infectiousness of fighting' (Richardson, 1946c: 130). By 'geography' he meant the 'opportunity of war for each country, depending on whether it was a worldwide sea power, a coastal state or a landlocked state'. By 'infectiousness' he meant the 'tendency to join the winning side.'[1]

Modern understandings of chaos and fractals were derived in the context of turbulence, which is different from a laminar flow where the volume of a fluid follows the same path as its predecessors. 'It is rather like the difference between the orderly progress of a well-disciplined company of soldiers and the wild rush of an unruly mob – the difference between order and chaos' (Ashford, 1985: 83). Chaos theory became prominent after Richardson's death, inspired by the Lorenz-Attractor in a simple weather model, and was extended to arms race models.

The complexity of Richardson's theory of atmospheric processes precluded manual weather forecasts, until the rise of computers after World War II. In 1946, John von Neumann proposed to the US Navy to apply high-speed, electronic, digital computing to dynamic meteorology. The ENIAC computer, derived for military purposes, was ready by March 1950 for the first test of simplified equations using meteorological observations, with promising potential to predict large-scale weather patterns (Ashford, 1985: 243). With the growing success of numerical weather prediction, Richardson's contributions became widely appreciated. To facilitate access to his annotated list of fatal quarrels, he presented a revised version

---

[1]Both quotations from Ashford (1985: 209).

in machine-readable form (Ashford, 1985: 257). Long before Geographical Information Systems and cellular automata, Richardson suggested cell-based geographical approaches to conflict analysis, together with other conceptions that later became successful.[2] In the following, some are highlighted for his arms race model which served as a starting point for the conflict model derived by the author and applied to environmental conflict.

8.2   The Framework of the Richardson Arms Race Model

8.2.1   Stability and Balance of Power

Richardson's study of conflict modelling was inspired by his meteorological work (Hess, 1995). Similar to weather forecasting, he tried to predict war by finding general laws, common to all nations. Following his empirical analysis of World War I, he derived a set of differential equations to describe the arms buildup between major powers in Europe during the 1930s, possibly leading to major war (Richardson, 1960a, b). Richardson's model is based on the assumption that for two countries each increases its own armament level proportional to the armament of an opponent (weighted by the defense coefficients) and reduces it proportional to its own armament (weighted by the fatigue coefficients) plus a grievance term.

The equilibrium where the armament levels of both sides do not change, corresponds to the so-called 'balance of power'. Its stability is determined by the eigenvalues of the matrix of coefficients: For a positive eigenvalue a deviation from the equilibrium grows exponentially (corresponding to instability), for a negative eigenvalue it decays asymptotically (indicating stability). For two nations instability is given if the product of their defense coefficients exceeds the product of their fatigue coefficients, indicating that the drivers of arms-buildup exceed the dampening factors. Then the arms race becomes unstable and escalates, while for stability the armament levels approach the equilibrium, favoring disarmament. Richardson extended the equations to several nations for the arms races 1909–14 and 1933–39, using military expenditures as armament variables, which was modified to the difference between threat and cooperation, taking into account beneficial relationships between nations, in trade, travel and correspondence (Richardson, 1938). These calculations supported his view that 'foreign policy had then a rather machine-like quality' (Richardson, 1960a: 33) and lead him to conclude that increasing armaments could lead to war breaking out, while a constant level of armament corresponds to a steady state without war.

---

[2]Cf. Gleditsch & Weidmann (2020) in this volume.

8.2.2   Critical Issues and Decision Rules

Richardson's model initiated a flood of publications on the armament dynamics and a debate about its applicability to real-world phenomena which raised several critical issues (see Smith, 2020, in this volume). Richardson himself was aware of the strengths and weaknesses of applying mathematics to social phenomena. Describing countries as structureless entities by one single variable was seen as questionable. Data on expenditure are not easily available and do not directly indicate security impacts. An arms race does not only have quantitative features, but also qualitative aspects, such as perceptions and doctrines. The Richardson model describes politics without personalities, where state authorities are black boxes and decisions are hidden in the budget. The fixed Richardson coefficients represent a linear and mechanistic interaction, where the initial conditions and coefficients determine the future, leaving no room for political decisions or control. Nations are assumed to have complete knowledge of the armament levels and react instanta-neously. In reality, each side has limited information about other countries, and worst-case assumptions provoke reactions towards arms buildup, often with deci-sion time lags.

The linearity and simplicity of the Richardson equations represents a few types of system behavior (oscillations, asymptotic decay, exponential increase). Reactions of real systems may be disproportionate and non-linear, showing qualitatively different modes of behavior. Decision-making may be better represented by time-discrete difference equations than by time-continuous differential equations. The arms buildup is not only an action-reaction process, driven by the opponents' armaments, but is also stimulated by a bureaucratic and budgetary dynamics with competing domestic interests. Although an arms race may provoke crisis-unstable situations, it does not necessarily lead to war if both sides want to avoid war or because one side reaches upper limits of armament which excludes an unlimited arms race.

Several extensions have been proposed to address the deficiencies of the Richardson model. Intriligator (1975) developed a framework for the strategic armament dynamics, based on decision rules that bridge the gap between desired and actual levels of missiles, taking into consideration the outcomes of a missile duel and the boundaries between deterrence and war initiation. This is represented by linear Richardson-type equations, whose coefficients can be derived from strategic considerations.

8.2.3   Chaos and Predictability in the Arms Race

While the Richardson model identifies basic system variables and relationships among countries, its rather simple structure does not represent the complexity of reality. Contrary to the well-ordered world of Newtonian mechanics, symbolized by the predictable swinging pendulum or the regular movement of the celestial bodies, complex systems, such as the turbulent weather patterns that Richardson studied, tend to be unpredictable. Since the 1970s, the natural sciences have begun to systematically explore critical phenomena, such as self-organization, tipping points, discontinuous phase transitions, and irreversibility. During the 1980s new mathematical concepts were developed, such as complexity, chaos, and non-linear dynamics. Chaos became not only a paradigm for the complex atmospheric dynamics but also for the turbulent transformation of the international system leading to the end of the Cold War in 1989 and the time after. Given the complexity of conflict, it seems appropriate to expand the Richardson model to non-linear phenomena.

The concept of chaos in arms race and war was introduced to show that simple non-linear deterministic arms race models may lead to the breakdown of predictability (Saperstein, 1984: 303). In chaos-like conflict situations, human actions and interactions are hardly foreseeable. Saperstein used a pair of non-linear difference equations with quadratic mappings for two variables, denoting the fractions of the available resources devoted to armaments which two countries pay annually. The problem of chaotic dynamics in arms race models was further investigated by Grossmann & Mayer-Kress (1989). The difference equations have a Richardson-like form with reaction parameters corresponding to the defense and fatigue coefficients and a grievance term, but with discrete time and a non-linear term that dampens armament expenditures at the upper cost limits. Factors provoking chaotic behavior are overshooting or underestimation, hectic responses, delay in information processing or discretization. The authors distinguish between chaos and instability: 'it is wrong to identify the general onset of bounded chaos with the outbreak of a war or another global crisis. The really dangerous case is instability' (Grossmann & Mayer-Kress, 1989: 702). Another non-linear time-discrete model, using decision rules for weapons procurement, was used by Saperstein & Mayer-Kress (1988) to simulate the impact of missile defense systems (Strategic Defense Initiative) on the East-West arms race. If production rates strongly increased, a chaotic transition from offense to defense occurred.

8.3 Multi-agent Interaction of Conflict and Cooperation

The Richardson model and other arms race models can be embedded into a broader framework of dynamic conflict modelling, bridging the gap between models for a few agents who optimize game strategies, and models for a large number of agents following dynamic decision rules. Inspired by Richardson's thinking about connections between the natural and the social world, in the following a dynamic agent-based modeling approach of social interaction is introduced that combines motivation and opportunity of multiple human agents to act upon and interact with their natural and social environment (for an overview see Scheffran & Hannon, 2007; BenDor & Scheffran, 2019).

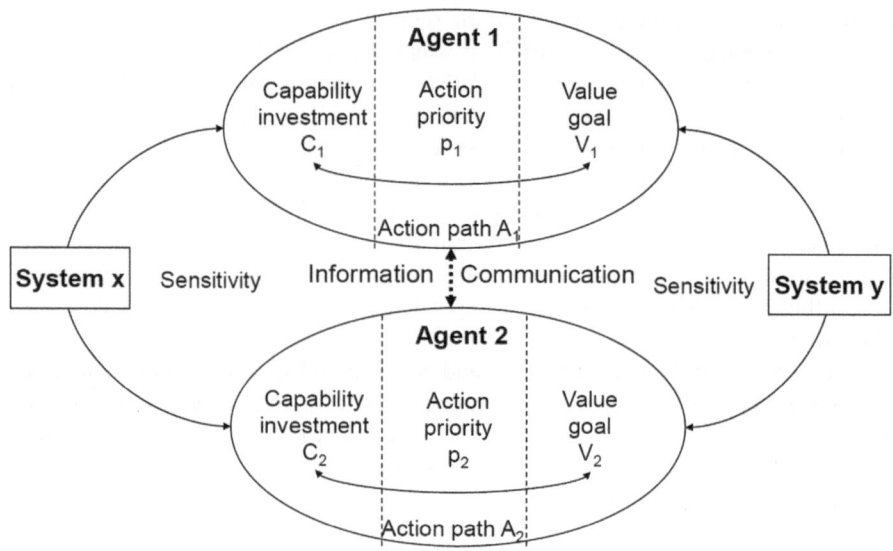

Fig. 8.1 Interaction between two agents and two environmental systems. Adapted from Scheffran et al. (2012)

8.3.1 From Individual Action to Multi-agent Interaction

The social interaction model follows the basic logic that individual agents act upon the environment, taking the opportunity to invest their capabilities to action pathways for achieving value-based goals which are a function of the benefits, costs, and risks of the actions taken. In repeated time steps and learning cycles agents mutually adapt their capabilities, action priorities and values, as a function of unit costs and values that represent the mutual sensitivities between agents and the environment (Figs. 8.1 and 8.2). Capabilities can change as a result of the dynamic interaction. Analytical conditions for conflict and cooperation as well as stability and chaos have been identified as a function of the value-cost efficiencies of each other's actions.[3]

Within the available capability limits, agents can adjust their investments and action pathways to meet their value goals according to decision rules, over time

---

[3]In more formal terms, the VCX model describes the dynamic action and interaction of agents who use part of their available capabilities (K) as investments (C) with priorities (p) to given action pathways (A) that change their environment (X). The observed impacts of actions are evaluated in each time step based on the agent's values (V) and goals (V*). Important parameters are the sensitivity of human value to environmental change (v_x) and the inverse sensitivity (unit cost) of environmental change to human investment (c_x). The respective value-cost ratio $f = v_x/c_x$ indicates how sensitive human value is to human investment and thus how efficient an action is. Negative efficiencies f indicate a conflicting action path where agents hurt their values.

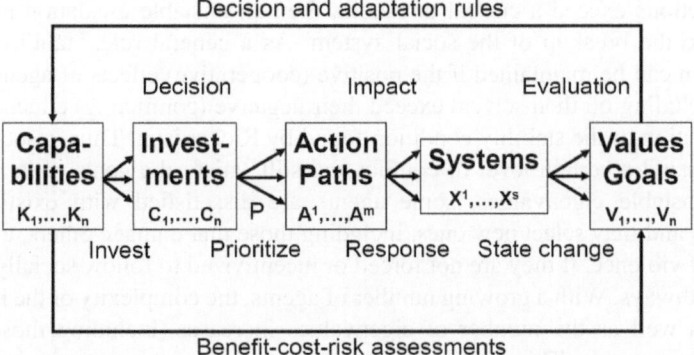

Fig. 8.2 VCX interaction model of multiple agents. Adapted from BenDor & Scheffran (2019)

switching to more efficient action pathways in a transition process. An interaction evolves for their responses to each other according to their respective decision rules which determine the fixed points where agents jointly achieve their goals. Decisive are the existence and location of the equilibria where all agents meet their goals within capability limits (corresponding to the 'balance of power' in Richardson's terminology) and the stability conditions of the multi-agent interaction matrix composed of the mutual action efficiencies.

Agents can control and stabilize or destabilize the dynamic interaction to some degree by using their capabilities and changing their action priorities to achieve the respective value goals. If the action priorities are directed towards hostile relations (damaging the values of other agents), the equilibria move towards higher investments, corresponding to an escalation. On the other hand, agents can switch to mutually beneficial cooperative actions lowering necessary investments. While this can be done independently, agents can also negotiate on their action priorities, leading to intertemporal dynamic games. In addition, they can form coalitions by pooling some of their invested capabilities and redistributing the gains (or losses) to the individual agents, or they may agree on the same values and goals, thus moving from individual to collective or institutionalized action and interaction (Scheffran, 2006).

8.3.2 Conditions for Stability

The type of social interaction is represented by the interaction matrix and its stability, mathematically determined by the eigenvalues around the social equilibrium. If agents are powerful in terms of their capabilities and efficient in pursuing their action goals, they can withstand, compensate, or counter-act a certain level of hostility by others, thus keeping eigenvalues in the negative range and avoiding major deviations from the equilibrium conditions. If the number and intensity of hostile actions exceed a critical threshold, then an unstable escalation may occur, leading to the breakup of the social system. As a general rule, stability of social interaction can be maintained if the positive (cooperative) effects of agents on each other (including on themselves) exceed their negative (conflicting) effects. This is a generalization of the stability condition found by Richardson. Thus, a social system can withstand a certain level of conflict and still satisfy the goals of its members.

For unstable eigenvalues, some agents are dissatisfied with existing action pathways and may select new ones, including those that damage others, e.g. though the use of violence, if they are not forced or incentivized to follow socially accepted action pathways. With a growing number of agents, the complexity of the interaction matrix as well as the number of eigenvalues increases, including those that are potentially unstable. This is known in systems theory as the 'complexity-stability' tradeoff, and raises the question whether complex systems are more unstable (Scheffran & Hannon, 2007; Gravel et al., 2016). Beyond tipping points complex systems may become destabilized and break apart into simpler ones, through cascading events and escalating conflict. Alternatively, mutual adaptations of actions or institutional control mechanisms can stabilize the interaction and contain conflict, e.g. by social security or other forms of support for the disadvantaged.

8.3.3 Connection to the Richardson Model

The presented multi-agent interaction model serves as a framework for the Richardson arms race model where countries are the main agents and their military expenditures are the investments. These are adapted according to decision rules driven by goal functions which are the differences between an adversary's expenditures and one's own (weighted by the defense and fatigue coefficients, serving as efficiencies) plus the grievance terms with strategic considerations. Both the Richardson model and the interaction model have a balance of power equilibrium and stability conditions for the interaction matrix when dampening coefficients exceed threatening ones. Richardson's focus on two countries is compatible with his observation that multi-country encounters tend to be more unlikely, unstable and chaos-like, corresponding to the complexity-stability tradeoff. He was thinking about the effect of alliances and organizations in multi-country contexts (Richardson, 1946c). Thus, Richardson has presented a role model with key elements relevant in the general model of social interaction.

8.4   Model Applications

The described interaction model has been applied to different fields, including arms races and arms control, economic production and environmental sustainability, resource conflicts, energy security and climate change (see Scheffran & Hannon, 2007; BenDor & Scheffran, 2019). To demonstrate its relevance, a few cases are selected beyond Richardson's narrower focus, including issues where atmospheric processes, weather patterns, and anthropogenic climate change could affect emerging conflict landscapes.

8.4.1   Complex Conflict Landscapes and the Spiral of Violence

The VCX model was born in the final phase of the Cold War, as part of the author's Ph.D. thesis in physics, to understand the stability of the nuclear arms race between the two major rivals. The study analyzed potential transitions from nuclear deterrence to a world where the nuclear threat is contained through missile defense (as suggested by former US President Ronald Reagan) or abandoned through nuclear disarmament (proposed by then Soviet General Secretary Mikhail Gorbachev). Simulating a shift from worst-case to best-case perceptions, or from hostile to friendly attitudes, the model showed chaos-like events beyond a tipping point, leading to nuclear disarmament (Scheffran, 1989). Shortly after, the Cold War ended in a domino effect and a breakup of the Eastern Block (Scheffran, 2008).

The following globalized world was characterized by growing connectivity between multiple agents and security dimensions. Within the model, the fractal security landscape was represented by a bifurcation diagram, which for increasing response rates of agents moves from periodic oscillations via stable equilibria to a sequence of multiple fixed points (Scheffran, 2003), challenging the predictability beyond the edge of chaos. Adding to complexity is an unstable interaction of multiple agents, some of whom benefit from the interaction while others suffer, separating into groups of 'winners' and 'losers' (Scheffran, 2003).

Some conflicts are related to the 'security dilemma', where threats to the security of agents provoke reactions that threaten the security of others. Key lessons can be drawn from the study of World War I and the diffusion of threats leading to it, using social network analysis to understand the arms race among alliances before this war (Vasquez et al., 2011), as studied in Richardson (1938). Beyond critical thresholds of instability, violent acts provoke more violent acts, leading to a self-enforcing 'spiral of violence', which today can be found in fragile regions of Africa. A key question is how to induce a transition to a self-enforcing cycle of cooperation and peace-building, similar to Richardson's change from threat to cooperation as part of 'collective security' (Richardson, 1935).

8.4.2   Climate Change, Social Instability, and Violent Conflict

Richardson was aware that weather and climate are among the most complex systems. Although the climate system has been largely stable in the Holocene, human interference may become a destabilizing factor in the Anthropocene if critical tipping points are exceeded (Steffen et al., 2018). Weather extremes such as hurricanes, droughts, forest fires, floods, and heatwaves, often correspond to non-linear mechanisms such as phase transitions, critical thresholds, and chaos. Natural disasters are generally associated with extreme consequences that burden the stability of natural and social systems and overwhelm their adaptive capacities. The effects may be aggravated by compound events, i.e. the complex combination of multiple climate drivers and hazards. Together, they are more likely and risky than their independent occurrence, e.g. concurrent hot and dry summers (Zscheischler et al., 2018).

In this context, climate change has been called a potential risk multiplier that combines with other risks, including those to human security and societal instability (such as forced displacement, riots, insurgency, intervention, urban violence, and civil war). The implications of compounding risks have not been sufficiently addressed in climate-conflict research where some studies claim climate change to be a significant driver of violent conflict, while others find no clear causality. This deficit can be addressed by an 'agent-based approach to assess the interplay between capabilities and motivations for violence and the conditions for conflicting or cooperative interactions. ... In the most affected regions, the erosion of social order and state failure as well as already ongoing violent conflicts could be aggravated, leading to a spiral of violence that further dissolves societal structures' (Scheffran et al., 2014: 369). One compound effect is the double vulnerability to violence and environmental hazard: environmental change can make societies more vulnerable to violence which in turn can make societies more vulnerable to environmental change, leading to a trap from which escape is difficult (Scheffran et al., 2014: 375).

These theoretical considerations increasingly attract empirical research on conflict sensitivity to climate change (von Uexkull et al., 2016: 12391): 'Results from naive models common in previous research suggest that drought generally has little impact. However, context-sensitive models accounting for the groups' level of vulnerability reveal that drought can contribute to sustaining conflict, especially for agriculturally dependent groups and politically excluded groups in very poor countries. These results suggest a reciprocal nature – society interaction in which violent conflict and environmental shock constitute a vicious circle, each phenomenon increasing the group's vulnerability to the other.'

Within the described model of social interaction, climate change may affect the allocation of investment to conflict, by undermining resource productivity (e.g. of agricultural output) and diminishing efficiency of human capabilities, or by providing incentives for violent resource capture, leading to stronger hostile actions.

In both cases, this can undermine the achievement of human goals und trigger more investments fueling conflict. Some effects could act over long distances, for instance large migration movements, interventions or humanitarian aid in remote regions affected by violent conflict. In this way, climate change may act as a global connector, adding to globalization, communication, transportation, and other linkages. To stabilize climate-induced interactions, agents could move towards mutually beneficial solutions (win-win), e.g. by innovation, resource sharing, risk management, and transition from high-emission to low-emission pathways within the temperature goals of the Paris Treaty (BenDor & Scheffran, 2018: Ch. 9).

Modeling climate-related conflict is still in an early stage and Richardson's work can provide some guidance, although he did not explicitly discuss the linkages between weather/climate and conflict, besides the observation that wars in the north temperate zone have ordinarily begun in spring or summer (Richardson, 1960b: 129). However, he pointed out that the probability of encounter in conflict is affected by geography (opportunity of war) and infectiousness (tendency to join the winning side) which are related to capability and motivation in conflict interaction. Richardson (1946c) also noted several factors that are important for conflict connectivity: 'Aviation is now tending to put every nation into contact with every other' (147) … 'The more persistent contact, the more opportunity for quarrels.' (152) … 'the trouble begins with the existence of a world-wide controversy' (155).

Apparently climate change is one such 'world-wide controversy'. Considering Richardson's general observations of multiple encounters, climate change could result in an increasing number of multi-actor and multi-factor encounters and related conflicts, which are not independent but result in compound risks. Further discussing these linkages could build bridges between Richardson's work in meteorology and peace.

8.5   Summary and Outlook

Starting from the Richardson arms race model and possible extensions, an integrative model framework of social interaction was presented in order to analyze conflict and cooperation, instability, tipping points, and cascading risks as well as transition and transformation processes. To cope with destabilizing consequences, affected systems need to adapt to the changing circumstances. Adaptive mechanisms influence critical decision points and adjust actions along multiple causal chains to protect human security and move from conflict to cooperation. The goal is to avoid risky pathways and facilitate a sustainable transformation, coping with conflict and climate change, developing social structures, political strategies, and institutional mechanisms that avoid or minimize social conflict and instability. Model expansions may contribute to improved understanding and forecasting of climate change and violent conflict in a turbulent world, encountering Richardson's research in atmospheric sciences and in peace and conflict studies.

References

Ashford, Oliver M (ed) (1985) *Prophet or Professor? The Life and Work of Lewis Fry Richardson.* Bristol: Adam Hilger.

BenDor, Todd K & Jürgen Scheffran (2019) *Agent-based Modeling of Environmental Conflict and Cooperation.* Boca Raton, FL: CRC Press.

Gleditsch, Kristian Skrede & Nils B Weidmann (2020) From hand-counting to GIS: Richardson in the information age. Ch. 7 in this volume.

Gravel, Dominique; François Massol & Mathew A. Leibold (2016) Stability and complexity in model meta-ecosystems. *Nature Communications* 7(12457): 1–8.

Grossmann, Siegfried & Gottfried Mayer-Kress (1989) Chaos in the international arms race. *Nature* 337(23 Feb): 701–704.

Hess, George D (1995) An introduction to Lewis Fry Richardson and his mathematical theory of war and peace. *Conflict Management and Peace Science* 14(1): 77–113.

Intriligator, Michael D (1975) Strategic considerations in the Richardson model of arms races. *Journal of Political Economy* 83(2): 339–353.

Richardson, Lewis Fry (1935) Mathematical psychology of war. *Nature* 142(28 Dec): 1025.

Richardson, Lewis Fry (1938) The arms race of 1909–13. *Nature* 142(29 Oct): 792–793.

Richardson, Lewis Fry (1946a) Chaos, international and inter-molecular. *Nature* 4004(158): 135.

Richardson, Lewis Fry (1946b) The probability of encounters between gas molecules. *Proceedings of the Royal Society of London. Series A, Mathematical and Physical Science* 186(1007): 422–431.

Richardson, Lewis Fry (1946c) The number of nations on each side of a war. *Journal of the Royal Statistical Society* 109(2): 130–156.

Richardson, Lewis Fry (1956) Mathematics of war and foreign politics. Ch. 6 in: James R Newman (ed) *The World of Mathematics*, vol. 2. New York: Simon & Schuster, 1240–1253.

Richardson, Lewis Fry (1960a) *Arms and Insecurity.* Pittsburgh, PA: Boxwood.

Richardson, Lewis Fry (1960b) *Statistics of Deadly Quarrels.* Pittsburgh, PA: Boxwood.

Saperstein, Alvin M (1984) Chaos – A model for the outbreak of war. *Nature* 309(24 May): 303–305.

Saperstein, Alvin M & Gottfried Mayer-Kress (1988) A nonlinear dynamic model for the impact of SDI on the arms race. *Journal of Conflict Resolution* 32(4): 636–670.

Scheffran, Jürgen (1989) *Strategic Defense, Disarmament, and Stability – Modelling Arms Race Phenomena with Security and Costs under Political and Technical Uncertainties.* PhD Thesis, Marburg: Department of Physics, IAFA Report (9).

Scheffran, Jürgen (2003) Calculated security? Mathematical modelling of conflict and cooperation. Ch. 20 in: Bernhelm Booss-Bavnbek & Jens Høyrup (eds) *Mathematics and War.* Basel: Birkhäuser, 390–412.

Scheffran, Jürgen (2006) The formation of adaptive coalitions. In: Alain Haurie, Shegeo Muto, TES Raghavan & Leo Petrosjan (eds) *Advances in Dynamic Games. Applications to Economics, Management Science, Engineering, and Environmental Management.* Basel: Birkhäuser, 163–178.

Scheffran, Jürgen (2008) The complexity of security. *Complexity* 14(1): 13–21.

Scheffran, Jürgen & Bruce Hannon (2007) From complex conflicts to stable cooperation. *Complexity* 13(2): 78–91.

Scheffran, Jürgen; Peter Michael Link & Janpeter Schilling (2012) Theories and models of climate-security interaction. Ch. 5 in: Jürgen Scheffran, Michael Brzoska, Hans Günter Brauch, et al. (eds) *Climate Change, Human Security and Violent Conflict.* Berlin: Springer, 91–132.

Scheffran, Jürgen; Tobias Ide & Janpeter Schilling (2014) Violent climate or climate of violence? Concepts and relations with focus on Kenya and Sudan. *International Journal of Human Rights* 18(3): 369–390.

Smith, Ron (2020) The influence of the Richardson arms race model. Ch. 3 in this volume.

Steffen, Will; Johann Rockström, Katherine Richardson, et al. (2018) Trajectories of the earth system in the anthropocene. *PNAS* 115(33): 8252–8259.

Vasquez, John A; Paul F Diehl, Colin Flint, Jürgen Scheffran, Sang-Hyun Chi & Toby J Rider (2011) The conflictspace of cataclysm: The international system and the spread of war 1914– 1917. *Foreign Policy Analysis* 7(2): 143–168.
von Uexkull, Nina; Mihai Croicu, Hanne Fjelde & Halvard Buhaug (2016) Civil conflict sensitivity to growing-season drought. *PNAS* 113(44): 12391–12396.
Vulpiani, Angelo (2014) Lewis Fry Richardson: Scientist, visionary and pacifist. *Lettera Mathematica* 2(3): 121–128.
Zscheischler, Jakob; Seth Westra, Bart Hurk, et al. (2018) Future climate risk from compound events. *Nature Climate Change* 8(6): 469–477.

Jürgen Scheffran, b. 1957, Ph.D. in Physics (University of Marburg, 1989); Senior Researcher and Faculty Member, University of Illinois (2004–09); Professor of Geography, University of Hamburg (2009–), head of Research Group Climate Change and Security at CliSAP/CliCCS Clusters of Excellence; research fields: climate change, migration and security, rural-urban interaction, agent-based modelling, sustainable peace, juergen.scheffran@uni-hamburg.de

Chapter 9
When Lanchester Met Richardson: The Interaction of Warfare with Psychology

Niall MacKay

Abstract Simple dynamical systems, in the spirit of Richardson's arms race, can be used to investigate the core dynamics of various models of insurgent and multilateral war. This chapter describes two such models. The first combines Richardson's two-nation arms race with Lanchester's attrition model and Deitchman's guerrilla variant of it to create a model in which the typical long-term outcome is neither annihilation nor escalation but rather a stable fixed point, a stalemate. The scaling it implies for the force required to defeat an insurgency matches that which has been observed. The second model is of multilateral attritional war, in the spirit of Richardson's multinational arms race. We describe the case of three antagonists, whose objective is to win but, if they cannot win, to minimize their remaining opponents. In contrast to truels and triads in which the objective is survival, and the weakest actor often emerges in a position of surprising strength, here the outcome is mutual annihilation, unless one side can beat the others put together.

Much of the controversy about the value of Richardson's arms race models can be set within a wider discussion of the value of mathematical modelling generally. An excellent modern essay is Joshua Epstein's (2008) 'Why model?', whose killer point is that a mathematical model at least gives a hypothesis that can be analyzed and falsified, in contrast to a purely verbal argument which may be too slippery to be tractable. Epstein also notes that a central purpose of simple mathematical models is to illuminate 'core dynamics' – that is, in dynamical systems, the ways that growth, decay, cyclicity and finally chaos appear. Whether in the FitzHugh-Nagumo model of an excitable system such as the human heart, the Kermack-McKendrick ('SIR') model of the epidemiology of a rapidly-developing

The original version of this chapter was revised: The name "Gregory D. Hess" has been corrected to "George D. Hess" in Reference list. The correction to this chapter is available at https://doi.org/10.1007/978-3-030-31589-4_12

This chapter builds on MacKay (2015). Permission granted by the Operational Research Society to reprint material from that article.

infectious disease, the Lotka-Volterra model of predator-prey ecosystems or Alan Turing's model of pattern formation, an understanding of the essential dynamics is a necessary beginning and complement to more elaborate simulations. As so often, Lewis Fry Richardson himself put it well:

> Strange to say, it is to the advantage of realism that mathematicians customarily replace the actual world by various idealized models. For they choose models that can be analyzed with ease; and thus they are free to think about the resemblances or misfits between the model and the actual world. If, with a solemn feeling of the importance of things as they really are, we were to admit the irregularities of the actual world into the statement of our problems, we should in consequence have to attend to enormous elaborations of mathematics in the process of solution, whereby our attention would for a long time be distracted away from the actual world (Richardson, 1960: 169).

But there is a stronger argument for the importance of simple models, namely that they necessarily capture the central truths hidden within more complex simulations, which, it must be understood, refine rather than supersede the simpler models. Approximations to complex models exist, describe their core dynamics, and can be classified. The classic example, post-dating Richardson, is Thom's 'catastrophe theory' from the 1960s, to which an excellent introduction is given by Poston & Stewart (1978). This was heavily oversold at the time, but is nowadays undervalued for what it contains, a classification of all the ways that change can occur in dynamical systems. Indeed, it is no more than slightly simplistic to characterize catastrophe theory as a body of results about Taylor's theorem in higher dimensions – and Taylor's theorem, taught in school calculus classes, is no more than the writing down of the simplest approximation to any nonlinear behaviour: first a constant, then a linear variation, then a quadratic, and so on. In that light, Richardson's arms races are just the simplest, linear approximation to the dynamics of the tension between antagonism and fatigue. Of course, a dynamical system may not be the correct conception of such a situation, but to the extent to which it is so, Richardson's conclusions necessarily follow. As he himself said, 'All that can be proved by mathematics is that certain consequences follow from certain abstract hypotheses' (Richardson, 1960a: 145).

This chapter is based on two ideas: The first is a small conceit, an imagined meeting in the world of the intellect, of LFR's ideas with those of someone very different, the irascible and occasionally belligerent British engineer Frederick Lanchester, who wrote down the first models of attritional war, with their grim calculus of constant warfare and resulting annihilation (MacKay, 2015). The original motivation was the observation that rather similar core dynamics pervade various of the attempts in the literature to model attrition in insurgent warfare, in which the military forces are bound up with more nebulous, psychological variables. Many of these attempts can be encompassed in a combination of Richardson arms races with Lanchestrian attrition and its variants. The action is all in the scaling, in whether it is the attrition or the escalation which scales faster, and whether it does so asymmetrically in a consistent way.

The second is in the spirit of Richardson's generalization of his arms races from two to an arbitrary number of countries. His conclusion there was that 'the world will for most of the time be content with just enough stability' (Richardson, 1960: 183). I analyze general multilateral wars of attrition but present the situation here for just three antagonists. There is some history of modelling conflict among three actors, and the generalization of a pistol duel – the truel – throws up similar conclusions in most of its variants: when each actor's overriding objective is survival, it is often the weakest actor whose prospects are the best. Our conclusions, however, are in stark contrast: in Lanchestrian attrition, whether Square or Linear Law, with the objective of maximizing one's own numbers minus one's opponents' numbers at the end of the war, either one actor can beat the others put together, or mutual annihilation is the robust outcome.

9.1 Richardson's Arms Race Model

We begin by looking at the 'phase plane' of the Richardson arms race. For reasons which will become apparent we denote by S and R the variables x and y in his arms race model (cf. Smith, 2020, in this volume), here restricted to be positive. The phase plane then shows a field of arrows which denote the direction (but not the magnitude) of the flow of (S, R), and also shows the curves (for Richardson's model they are lines) on which the flow is either horizontal or vertical. These 'null clines' therefore intersect at points at which the flow is both horizontal and vertical: that is, where there is no flow, so that the intersections are the 'fixed points' of the system. Figure 9.1 shows the two qualitatively different possibilities. In Fig. 9.1a, on the left, fatigue outweighs antagonism. Whatever the starting point, the resulting flow reaches the stable fixed point, an equilibrium in which the two sides' forces are somewhat larger than would be the case in the absence of antagonism but are nonetheless stable. Figure 9.1b, on the right, illustrates what happens when antagonism increases: the null clines have passed through being parallel and now diverge, resulting, from all starting points, in a runaway arms race. I have heard it objected that there never has been an exponentially growing arms race of the kind implied, but of course the phase plot tells us nothing about the magnitude of the flow, and one can perfectly well nonlinearly rescale the time variable to alter the functional form of the rate of growth.

In (a) fatigue is predominating, resulting in a stable fixed point; in (b) antagonism is predominating, resulting in runaway arms growth. The null clines (of horizontal or vertical flow) are shown as thin red lines.

9.2 A Lanchester-Richardson Model

In Richardson's model the parameters which link the variables with their rates of change are psychological. A very different classic warfare model is that of Lanchester (1914), which describes attritional war. In Lanchester's 'aimed-fire'

(a)   **(b)**

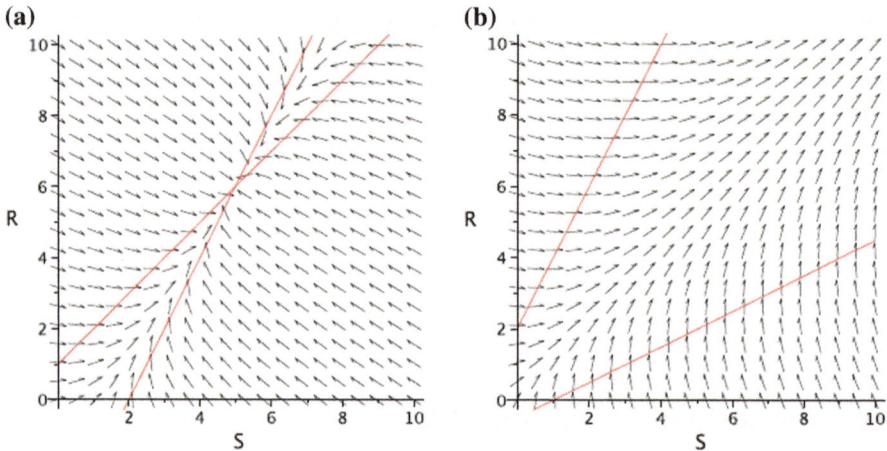

Fig. 9.1 The Richardson arms race

model, each force causes damage in proportion to its numbers, and the constant of proportionality is the rate at which each unit kills its enemies. There is little intrinsic interest in the phase plane, for there is no possibility of fixed points other than mutual annihilation $(S, R) = (0, 0)$. The classic result, rather, results from the trajectories' being hyperbolae: the difference between the forces' kill-rates multiplied by the square of their numbers is constant, and thus determines the winner, in Lanchester's Square Law. Mathematically, of course, this is similar to Richardson's model, with altered parameters (merely), not altered (linear) dynamics: fatigue and constant grievance are absent, and antagonism has a negative effect on opponents.

Suppose that we combine the two models. At its simplest, the model is simply as in Fig. 9.1, but with antagonism reduced by kill-rate. The growth of the forces is now supplemented by the horror of continuous attrition to create either a stable equilibrium, or (if antagonism still outweighs all else) an arms race continuing during open war. Both situations are perhaps not so different from that of 1914–17: the fixed point in Richardson's model can just as well describe stasis in 'hot' war as in 'cold'. Nowhere in either phase plane do we see either side 'winning'.

Now suppose instead that we make the model asymmetric between S and R, letting S be the 'state' and R the 'rebels' or 'revolution' in an insurgent war, and think about the conditions for a state win – that is, for the state to annihilate the rebels. Assume that the state uses its resources both to inflict high damage (in proportion to its own strength) on the rebels and to reduce the rate at which by its antagonism it causes the rebellion to grow, so that the former is greater than the latter. In contrast the state is able to reinforce itself in proportion to rebel numbers at a greater rate than it loses units to them. There are then two possible outcomes, whose controlling parameter is the (net negative) rate at which the state antagonises the rebellion minus the rate at which it kills the rebels. The result is, in Fig. 9.2a on the left, a low-antagonism regime which sees an inevitable state win, or, on the right in Fig. 9.2b, a high-antagonism regime which may see either (for sufficiently large

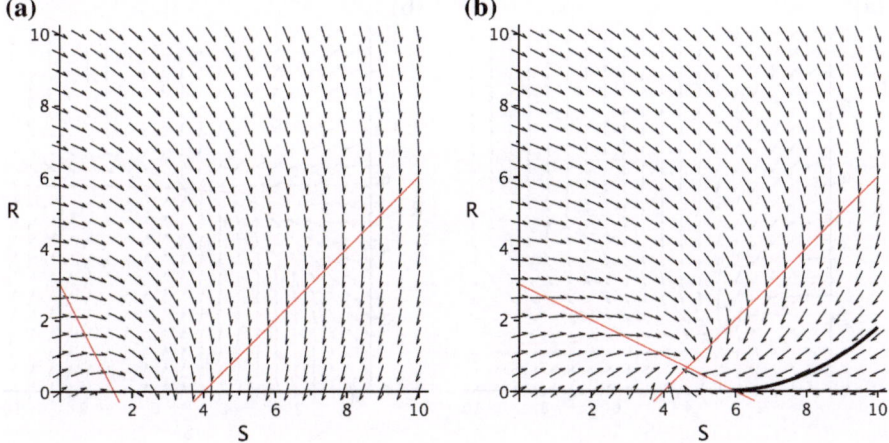

Fig. 9.2 A combined Lanchester-Richardson model. **a** Low antagonism, resulting in a state win, or **b** high antagonism, resulting either in a state win or – if the state is initially relatively weak – a fixed point (stalemate). The thin red lines are the null clines; the thick curve in (**b**) is the separatrix between stalemate and state win

initial S and small initial R, to the right of the thick curve) a win for the state, or (otherwise) a fixed point, a stalemate.

This model captures some useful features. Set up in this way, it does not allow a rebel takeover. But it does distinguish an ongoing insurgency from a state win, with the latter requiring a reasonable set of conditions: effective attrition of the insurgency, low antagonism by the state, and a relatively strong initial state position. But (and this goes back to the issue of 'core dynamics'), it probably gets one crucial ingredient wrong: insurgencies almost certainly do not have the (symmetric) Lanchester square-law model as their attritional dynamics.

9.3   A Deitchman-Richardson Model

It is a standard feature of asymmetric variants of Lanchester models intended to describe insurgent or guerrilla warfare that (at its simplest) R's fire is 'aimed' whereas S's is 'unaimed', that the rebellion is able to target its fire more efficiently that the state (Deitchman, 1962). The upshot is an asymmetry of scaling, in which the state's action becomes relatively more effective when the insurgency is larger. The horizontal-flow null cline is then not a straight line but a curve, and the phase portraits are as shown in Fig. 9.3. Here both low antagonism (left, Fig. 9.3a) and high antagonism (right, Fig. 9.3b) result in a stalemate. There can be no unlimited escalation, for the asymmetric nonlinear scaling of the attrition of the rebellion by the state prevents it.

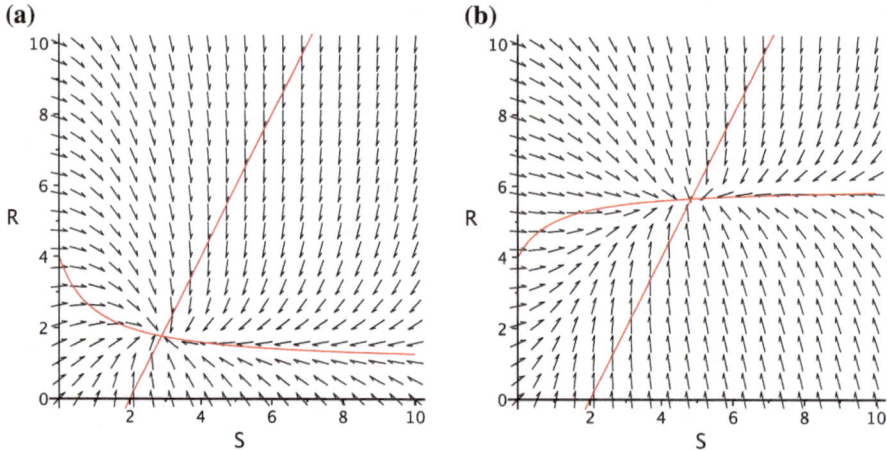

Fig. 9.3 A combined Deitchman-Richardson model has a stable fixed point for both **a** low antagonism and **b** high antagonism

In this Deitchman-Richardson model the only way for the state to win is for its ability to direct aimed fire – targeted, intelligent military action – at the rebellion to outweigh its antagonism. If this effect is sufficiently strong (Fig. 9.4a) the state always wins. If not (Fig. 9.4b) then there is, as in Fig. 9.2, a separatrix between the regime in which the state is sufficiently initially relatively strong to win and that in which a stalemate is reached.

These results are robust to more generalized scalings and subsume various results in the literature. Of course, a 2D dynamical system can do no more than illustrate some simplified dynamics and can anyway capture behaviour no more complex than growth, decay or a (perhaps cyclic) stale-mate. But the general conclusion is that simple Richardson-like linear antagonism, when combined with Lanchester-Deitchman theory in which the theoretical signature of insurgent warfare is that insurgent attrition scales faster than state attrition, typically results in stalemate.

This makes the crux of any empirical verification clear. Good time-series data on insurgent wars are hard to find: not only are the parties otherwise engaged, but the rebellion – both R and its growth rate dR/dt – remains largely hidden. But what one can do is to examine, for a set of insurgencies, the relationship between the level S of state military force needed to win and the effectiveness rate dS/dt of rebel military action. Goode (2009–10) did precisely this, analyzing 42 insurgencies. The theoretical expectation in Deitchman's model, deduced from its analogue of the Square Law, is that $S^2 \sim R$ (where $\sim$ is to be read as 'scales like'), while $\dot{S} \sim R$: combining these, $S \sim \sqrt{dS/dt}$ Goode found $S \sim (dS/dt)^{0.45}$, a strikingly good match.

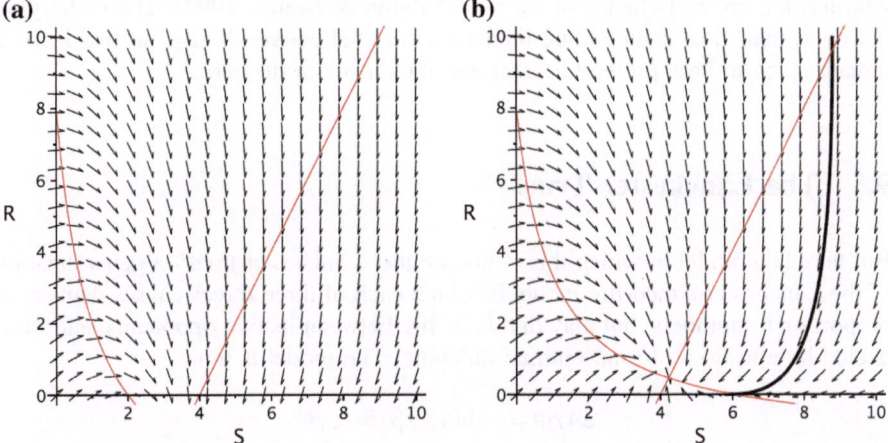

Fig. 9.4 Combined Deitchman-Richardson model with state targeted action greater than antagonism. In **a** when this effect is large, the state wins; in **b** when this effect is modest, the state wins if the initial rebellion is relatively small, else a stalemate is reached

9.4   Models with Three Actors

Richardson extended his model first to three and then to a larger general number of actors. In the case of three actors, suppose first that all individual actors have the same rate of fatigue, and the strengths of all pairwise antagonisms are equal. Pairwise, each interaction results in an arms race if and only if antagonism is greater than fatigue. Unfortunately, when we have three actors the situation is worse, for there is more antagonism going on – indeed it is, in a precise sense, doubled. An arms race occurs if and only if the pairwise antagonism is greater than half the individual rate of fatigue (Richardson, 1960: 154–155).

Now that we have three actors, however, we can introduce decision parameters and create differential games. We can also allow coalitions or alliances. A great deal of literature covers such possibilities, but a common thread runs through much of it: the apparent weakest actor is in many cases in the strongest position. For example, in a simple quantitative model of three forces, the largest force or coalition being the winner, it is typically the smallest force which has the greatest capacity to ensure it ends up on the winning side (Caplow, 1956: 490): 'the triadic situation often favors the weak over the strong'.

Battles among three actors, or 'truels', can be set up in many ways. Perhaps the very simplest is to give *A, B* and *C* each one shot and require them to act, in that order. What should *A* do? If he shoots *B*, then *C* will shoot him. If he shoots *C*, then *B* will shoot him. Yet suppose he shoots in the air: then, by making himself powerless, he has become no threat to *B*, who shoots *C*, leaving *A* and *B* standing. Many variations – deterministic or probabilistic, with shooting in fixed or in random order – are possible, and it is often the worst shot or otherwise-weakest actor who has the greatest chance of success (Kilgour & Brams, 1997). The underlying reason is that most actors value their own survival above all else, so that to be a threat is also to be a preferred target and thus to be in danger.

9.5   The Lanchester Truel

But here is a model in which this is not so: the 'Lanchester truel', a generalization of the Lanchester aimed-fire model in which each of three forces causes damage in proportion to numbers, but may divide its fire between its two opponents, and must decide how to do so for an optimal outcome. The model is thus

$$dA/dt = -b(1 - \beta)B - c\gamma C$$
$$dB/dt = -a\alpha A - c(1 - \gamma)C$$
$$dC/dt = -a(1 - \alpha)A - b\beta B$$

where $A(t)$, $B(t)$, $C(t)$ are the force numbers and a, b, c their individual kill-rates, all positive. The three forces begin with given numbers $A(0) = A_0$, $B(0) = B_0$, $C(0) = C_0$ units. Each unit of A kills opponents (of either type) at rate a, B does so at rate b, and C at rate c. When one force is eliminated the other two fight a Lanchester aimed-fire duel, and the truel finishes when only one force remains. We'll call these final values, only one of which can be positive, A_∞, B_∞ and C_∞.

All of the parameters described so far – the initial numbers A_0, B_0, C_0 and the kill-rates a, b, c – are fixed for a given truel. But each force also has a decision parameter under its (and no one else's) control: for A it is α, for B it is β, for C it is γ. If $\alpha = 1$ then A targets only B; if $\alpha = 0$ then A targets only C; and the two are interpolated by $0 \leq \alpha \leq 1$. Each side can vary its parameter continuously throughout the fight if it wishes.

A set of coupled linear differential equations such as these is straightforward to solve, but the solution in itself is not very instructive (Kress et al., 2018a). The crucial first step is to decide on the objective. This is a war of attrition, but there is no equivalent, for general values of the parameters, of Lanchester's Square Law for this model – technically, this is because there is no quadratic quantity which remains the same throughout the battle, at least for general values of the parameters. Instead we take A's objective to be to maximize

$$A_\infty - B_\infty - C_\infty$$

and likewise for B and C. That is, if A can win then it wishes to do so with the largest possible remaining force, but if it cannot win then it wishes to minimize the sum of its opponents' remaining forces. It does so without discriminating between B and C, but in fact this does not matter: the results apply to any objective function which is increasing in A_∞ and decreasing in B_∞ and C_∞. Thus, the outcome in the end does not depend on the implied relative worth of one's own survival against one's opponents' destruction.

This outcome, in contrast to most truels and coalition models, is that either one side can beat the other two put together, or all three actors are locked in an attritional stalemate which leads to mutual annihilation. The reason is the valuing of the destruction of opponents by an actor which cannot emerge victorious. The proof is in two parts, and, while I shall not reprise the mathematical details, the essential content is easy to describe with the aid of Fig. 9.5. Here, for simplicity, I have projected (A, B, C) onto the positive octant of the unit sphere, thereby concerning us only with the relative numbers of A, B and C. In reality, the dynamics is, of course, of absolute numbers of A, B and C always declining.

Consider just one corner of this octant, where A has very large numbers relative to B and C, and A dominates. That is to say, if A chooses the correct fire distribution between B and C, following the prescription of Lin & MacKay (2014), then it can

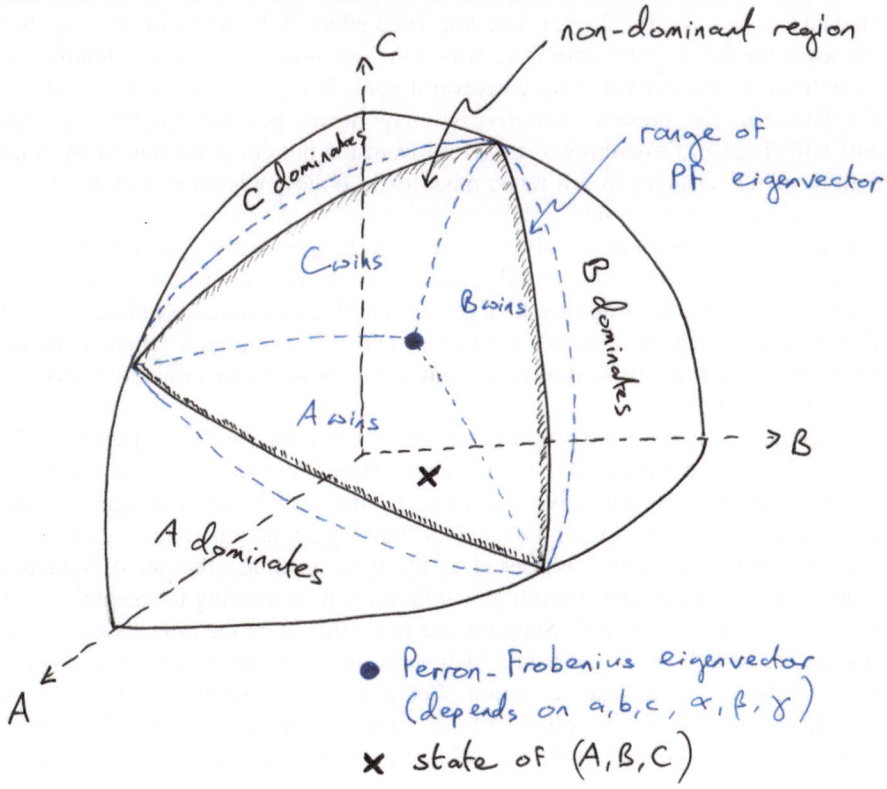

Fig. 9.5 The Lanchester truel. Force sizes **A**, **B**, **C** are projected onto the unit sphere. In the hachured triangle no one actor 'dominates' – that is, can beat the other two together. The game-theoretic outcome is that by altering their fire distributions the actors move the condition for collective mutual annihilation (dot) onto the present state (cross)

guarantee to beat the other two forces even if they each fire only at A. Likewise for B and C. These three regions bound (with the hachured lines) the central triangle, the non-dominant region, within which none of A, B or C can by their own choices guarantee to win.

Now consider the markings, which describe a lovely result in linear algebra. The dot is the Perron-Frobenius (or PF) eigenvector, the state from which the dynamics send (A, B, C) directly in a straight line to the origin (thereby remaining static in the projection onto the unit sphere), to collective annihilation. Its position depends on the actors' policies α, β, γ, and moves as these policies change. Within what bounds can it move? These are shown as the dashed lines, and it is an elegant result that the range of the dot, the PF eigenvector, bounded by the dashed lines, strictly includes the non-dominant region, with equality when $a = b = c$.

The rest of the result is a straightforward differential game. The dot is a Nash equilibrium, a position from which, within the non-dominating region, none of A, B or C can do better (according to their objectives) by departure; and the combined action of A, B and C will always push the dot onto whatever is the current state (marked as a cross on the figure). The combined effect of the actors optimizing their policies is for the dot (the state from which the outcome is mutual annihilation) to chase the cross (the current state) wherever it goes. This gives the result a great deal of robustness: the present state (cross) may move, but the annihilating state (dot) will chase it. Departures of the cross from the dot could be caused by small reinforcements, changes in kill rates, missteps in policy; whatever. But as long as the cross does not move into a dominating region, and as long as the actors can change their policies more rapidly than attrition changes the state of the forces, the dot will catch up with the cross and remain there. In fact, the result is much more general even than this: it applies to a general number of actors, to kill-rates which depend on opponent as well as on attacker, and to Lanchester's 'ancient' (linear law) model of warfare as well as to his aimed fire (Square Law) model (Kress, Lin & MacKay, 2018b).

Where does this leave the original truel insight, that the weakest is paradoxically strong? Can this be replicated in the Lanchester truel? In fact, it can, but one needs to choose a different objective function. In the Lanchester one-against-many problem, the optimal policy can be fixed by thinking about quadratic quantities, but it is identical with the policy produced by trying to maximize the rate of reduction of one's own casualty rate. Here this would amount to A trying to maximize d^2A/dt^2, and likewise for B and C. Suppose one takes this to be the objective. One can then impose a differential game in which each side alters its policy so as to maximize its own casualty-rate reduction. Let $a > b > c$, so that A is the most individually-effective force, followed by B, then C. It turns out that if A's and B's numbers are approximately balanced, C can judiciously divide its fire between A *and B* and emerge the winner. Why do A and B not also fire at C? Essentially because A is the bigger threat to B, and vice versa: and derogation by A and B from concentrating on each other, by firing at C, will only make their casualties worse. Within the cube $0 \leq \alpha$, β, $\gamma \leq 1$ of fire policies, this is the outcome with the largest basin of attraction; there is a smaller line-segment for which the same is true of an *A-C* duel with *B* looking on; and, only for $b + c > a$, a small line-segment of stable fixed points of *B-C* duel with *A* looking on.

The action, then, is all in the objective function: the overall conclusion is that, in multilateral war, if actors value their opponents' destruction as well as their own survival, then truel-like results no longer hold: rather, if no one actor can beat all the others put together, then all of them are headed for mutual destruction. If, instead, actors value only their own hurt, then something like classic truel results may follow. One can also show that, if your opponents value your destruction but you do not value theirs, then you must also inflict hurt on them (even if your objective is only to avoid your own hurt); it does no good with such opponents to withhold any of your own fire.

9.6   Concluding Thoughts

A fine review of Richardson's arms race models, including the multi-actor case and connecting it to the modern theory of dynamical systems and properties of the associated matrices, is given by Hess (1995). The recent literature on the modelling of insurgent war, much of which falls under the scope of generalized Lanchester-Richardson models (MacKay, 2015), still has much to learn from Richardson's work. Indeed, there are many more connections to be made in work on multilateral conflict, not only between arms races and attritional war, but also with the literature on stability in complex ecosystems which has grown from the pioneering work of May (1973), and even, in ongoing work which brings attention full-circle back to pressing questions of human social structures, with the stability of banking systems (Haldane & May, 2011; Bardoscia et al., 2017).

References

Bardoscia, Marco; Stefano Battiston, Fabio Caccioli & Guido Caldarelli (2017) Pathways towards instability in financial networks. *Nature Communications* 8: 14416.

Caplow, Theodore (1956) A theory of coalitions in the triad. *American Sociological Review* 21(4): 489–493.

Deitchman, Seymour (1962) A Lanchester model of guerrilla warfare. *Operations Research* 10(6): 818–827.

Epstein, Joshua (2008) Why model? *Journal of Artificial Societies and Social Simulation* 11(4): 12.

Goode, Steven M (2009–10) A historical basis for force requirements in counterinsurgency. *Parameters* (Winter): 45–57.

Haldane, Andrew & Robert May (2011) Systemic risk in banking ecosystems. *Nature* 469(7330): 351.

Hess, George D (1995) An introduction to Lewis Fry Richardson and his mathematical theory of war and peace. *Conflict Management and Peace Science* 14(1): 77–113.

Kilgour, D Marc & Steven J Brams (1997) The truel. *Mathematics Magazine* 70(5): 315–326.

Kress, Moshe; Jonathan Caulkins, Gustav Feichtinger, Dieter Grass & Andrea Seidl (2018a) Lanchester model for three-way combat. *European Journal of Operational Research* 264(1): 46–54.

Kress, Moshe; Kyle Lin & Niall MacKay (2018b) The attrition dynamics of multilateral war. *Operations Research* 66(4): 950–956.

Lanchester, Frederick (1914) *Aircraft in Warfare: The Dawn of the Fourth Arm.* London: Constable. [Based on 1913–14 articles in *Engineering* 98: 422–423 and 452–453. The chapter which deals with the mathematical models is reprinted in James Newman (ed.) (1956) *The World of Mathematics* 4: 2138–2157. New York: Simon and Schuster, reprinted by New York: Dover, 2000.]

Lin, Kyle & Niall MacKay (2014) The optimal policy for the one-against-many heterogeneous Lanchester model. *Operations Research Letters* 42: 473–477.

MacKay, Niall (2015) When Lanchester met Richardson, the outcome was stalemate: A parable for mathematical models of insurgency. *Journal of the Operational Research Society* 66: 191–201. [Reprinted as Ch. 6 of RA Forder (ed.) (2015) *Operational Research, Defence and Security.* Operational Research Essentials. London: Palgrave Macmillan, in association with the Operational Research Society: 124–147.]

May, Robert (1973) *Stability and Complexity in Model Ecosystems.* Princeton, NJ: Princeton University Press.

Poston, Tim & Ian Stewart (1978) *Catastrophe Theory and Its Applications.* London: Pitman.

Richardson, Lewis F (1960) *Arms and Insecurity: A Mathematical Study of the Causes and Origins of War.* Pittsburgh, PA: Boxwood.

Smith, Ron P (2020) The influence of the Richardson arms race model. Ch. 3 in this volume.

Niall MacKay, b.1967, B.A. in Mathematics (Cambridge, 1988), Ph.D. in Theoretical Physics (Durham, 1992); Professor and Head of Department of Mathematics, University of York; publications in mathematical physics, operations research and history, niall.mackay@york.ac.uk.

Chapter 10
On the Frequency and Severity of Interstate Wars

Aaron Clauset

Abstract Lewis Fry Richardson argued that the frequency and severity of deadly conflicts of all kinds, from homicides to interstate wars and everything in between, followed universal statistical patterns: their frequency followed a simple Poisson arrival process and their severity followed a simple power-law distribution. Although his methods and data in the mid-20th century were neither rigorous nor comprehensive, his insights about violent conflicts have endured. In this chapter, using modern statistical methods and data, I show that Richardson's original claims are largely correct, with a few caveats. These facts place important constraints on our understanding of the underlying mechanisms that produce individual wars and periods of peace and shed light on the persistent debate about trends in conflict.

10.1 Introduction

Lewis Fry Richardson (1881–1953) stands as one of the founding fathers of the modern field of complexity science (Mitchell, 2011), which aims to understand both how complexity arises from the interaction of simple rules and how structure emerges from the chaos of contingency. One of his most celebrated works was his analysis of the frequency and severity of interstate wars and other deadly conflicts (Richardson, 1944, 1948, 1960). Richardson also played critical roles in two other major pieces of complexity science, which continue to inform scientific efforts to understand systems as varied as developmental biology, the formation of galaxies, and the collective behavior of humans in its many forms.

The first of these arose in his work on the 'coastline paradox', which is captured by a deceptively simple question: how long is the British coastline? Richardson showed that the length of a coastline depends, paradoxically, on the length of the ruler used to measure it – the shorter the ruler, the longer the coastline's total length. This effect, now called the Richardson effect, paved the way for Benoit Mandelbrot's celebrated work on fractal geometry (Mandelbrot, 1967), which has informed numerous studies of complex social, biological, and technological systems (Mitchell, 2011). Richardson's insight also foreshadowed his discovery of a 'scale-free' pattern in the statistics of wars.

© The Author(s) 2020

N. P. Gleditsch (ed.), *Lewis Fry Richardson: His Intellectual Legacy and Influence in the Social Sciences*, Pioneers in Arts, Humanities, Science, Engineering, Practice 27, https://doi.org/10.1007/978-3-030-31589-4_10

The second arose from Richardson's pioneering work in meteorology, which was his primary focus for many decades. Much of his work here aimed to develop the mathematics of weather forecasting, and in recognition of those contributions, a dimensionless quantity related to buoyancy and shear flows in turbulent systems is called the Richardson number. Richardson (1922) also pioneered the use of numerical approaches to forecast the weather, despite the fact that sufficient computing power to make useful weather predictions would not be developed until several decades later. In this way, Richardson very nearly discovered, almost a half century earlier, the same mathematical chaos lurking in the equations of turbulence that Lorenz (1963) would later make world famous. Richardson's work on weather forecasting also foreshadowed his interest in the long-term statistics of wars.

These intellectual threads came together in Richardson's foundational studies of violent conflict, in which he argued that the frequency and severity of deadly 'quarrels' of all kinds, from small-scale events like homicides to large-scale events like interstate wars, followed universal statistical patterns (Richardson, 1960). Although little attention is now paid to his claims about small-scale events like homicides, Richardson's ideas about larger events have become central to the study of political conflict, including civil unrest (Biggs, 2005), terrorism (Clauset et al., 2007), insurgency (Bohorquez et al., 2009), civil wars (Cederman, 2003; Lacina, 2006), and interstate wars (Cederman, 2003; Cederman et al., 2011; Pinker, 2012; Harrison & Wolf, 2012). In this chapter, I focus on Richardson's ideas about the statistics of interstate wars.

Richardson's original analysis only covered interstate wars from 1820–1945 (Richardson, 1948). On the basis of these events, he made two claims about their statistical pattern. First, he argued that the sizes, or 'severities,' of these wars followed a precise pattern, called a power-law distribution, in which the probability that a war kills x people is $\Pr(x) \propto x^{-\alpha}$, for all $x \geq x_{\min} > 0$, and where $\alpha > 1$ is called the 'scaling' parameter. Second, he argued that the timing of wars followed a simple Poisson process, implying both a constant annual probability for a new war and a simple geometric distribution for years between wars (Richardson, 1944).

Although his statistical methods were not rigorous by modern standards and his data were far less comprehensive, these patterns – a power-law distribution for war sizes and a Poisson process for their onsets – represent a simple and testable model for the statistics of interstate wars worldwide. Crucially, Richardson's model is 'stationary,' meaning that the rules of generating new wars do not change over time.

If the empirical statistics of interstate wars really do follow the simple patterns claimed by Richardson, it would indicate strong constraints on the long-term dynamics of the underlying social and political mechanisms that generate wars and periods of peace (Ray, 1998; Ward et al., 2007; Leeds, 2003; Jackson & Nei, 2015; Alesina & Spolaore, 1997). A long-running debate within the study of conflict has focused on whether or not such conflicts are characterized by genuine trends (see Gleditsch & Clauset, 2018 for a recent review).

If the underlying mechanisms that produce wars are stationary, then any 'trend' is inherently illusory. However, deciding whether trends exist has proved difficult to resolve, in part because there are multiple ways to answer this question, depending

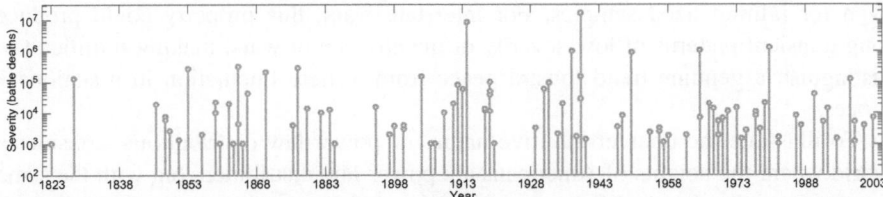

Fig. 10.1 Interstate war 1823–2003. The graph shows severity (battle deaths) and onset year for the 95 conflicts in the 181-year period based on data from the Correlates of War (CoW) interstate war data (Sarkees & Wayman, 2010). The absolute sizes of wars range from 1,000, a minimum by definition, to 16,634,907, the number of recorded battle deaths in the Second World War. Delays between consecutive war onsets range from 0 to 18 years, and average 1.91 years. Most wars (79%) ended no more than two years after their onset. Originally published in Clauset (2018)

on what type of conflict is chosen, what variable is analyzed, and how the notion of change is formalized. Different choices can lead to opposite conclusions about the existence or direction of change in the statistics of conflict (Payne, 2004; Harrison & Wolf, 2012; Braumoeller, 2013; Cirillo & Taleb, 2015; Gleditsch & Clauset, 2018; Clauset, 2018).

Here, I consider a more straightforward question: given modern statistical tools and interstate war data, do Richardson's claims about statistical patterns hold up, and if so, what does that imply about the long peace of the post-war period? For this investigation, I apply state-of-the-art methods (Clauset et al., 2009; Clauset & Woodard, 2013) to the set of interstate wars 1823–2003 given in the Correlates of War data set (Sarkees & Wayman, 2010) (Fig. 10.1). This data set provides comprehensive coverage in this period, with few artifacts and relatively low measurement bias, and allows us to focus on a period during which Richardson's model is plausible (Cederman et al., 2011).

10.2   Preliminaries

Before analyzing any data, I must clarify several epistemological issues and the impact of different assumptions on the accuracy and interpretation of the analysis.

Power-law distributions have unusual mathematical properties (Newman, 2005), which can require specialized statistical tools to analyze. (For primers on power-law distributions in conflict, see Cederman, 2003; Clauset et al., 2007.) For instance, when observations are generated by a power law, time series of summary statistics like the mean or variance can exhibit long fluctuations resembling a trend. The largest and longest fluctuations occur for a scaling parameter of $\alpha < 3$, when one or both the mean and variance are mathematically infinite, i.e., they never converge, even for infinite-sized samples. For interstate wars, this property could produce long transient patterns of low-severity or the absence of wars, making it difficult to distinguish a genuine trend toward peace from a mere fluctuation in a stationary process.

To illustrate the counter-intuitive nature of power-law distributions, consider a world where the heights of Americans are power-law distributed, but with the same average as reality (about 1.7 m), and I line them up in a random order. In this world, nearly 60,000 Americans would be as tall as the tallest adult male on record (2.72 m), 10,000 individuals would be as tall as an adult male giraffe, one would be as tall as the Empire State Building (381 m), and 180 million diminutive individuals would stand only 17 cm tall. As we run down the line of people, we would repeatedly observe long runs of relatively short heights, one after another, and then, rarely, we would encounter a person so astoundingly tall that their singular presence would dramatically shift our estimate of the average or variance of all heights. This is the kind of pattern that we see in the sizes of wars.

Identifying a power-law distribution within an empirical quantity can suggest the presence of exotic underlying mechanisms, including nonlinearities, feedback loops, and network effects (Newman, 2005), although not always (Reed & Hughes, 2002), and power laws are believed to occur broadly in complex social, techno-logical, and biological systems (Clauset et al., 2009). For instance, the intensities or sizes of many natural disasters, such as earthquakes, forest fires, and floods (Newman, 2005), as well as many social disasters, like riots and terrorist attacks (Biggs, 2005; Clauset et al., 2007), are well-described by power laws.

Testing if some quantity does or does not follow a power law requires spe-cialized statistical tools (Resnick, 2006; Clauset et al., 2009), because uncertainty tends to be greatest in the upper tail, which governs the frequency of the largest and rarest events, i.e., the frequency of large wars. Modern statistical tools provide rigorous methods for estimating and testing power-law models, distinguishing them from other 'heavy-tailed' distributions, and even using them to make statistical forecasts of future events (Clauset & Woodard, 2013).

Poisson processes pose fewer statistical issues than power-law distributions. However, for consistency, our analysis applies similar methods to both war timing and size data. Specifically, I estimate an ensemble of models, each fitted to a bootstrap of the empirical data, which better represents our statistical uncertainty than would a single model. Technical details are described in Clauset (2018).

Richardson's models are defined in terms of absolute numbers, i.e., the number of interstate wars per year and the number of battle deaths per war. Hence, I consider war variables in their unnormalized forms and consider all recorded interstate wars, meaning that our analysis takes the entire world as a system.

Analyses of interstate war statistics sometimes normalize either the number of wars or their sizes by some kind of reference population, e.g., dividing a war's size by the global population at the time. Such normalizations represent additional theoretical assumptions about the underlying data generating process.

For instance, normalizing the number of wars per year by the number of pairs of nations that could be at war assumes that war is a dyadic event and that dyads independently generate conflicts with equal probability (Ward et al., 2007). This choice of normalizer grows quadratically with the number of nations and will create the appearance of a trend toward fewer wars, even if Richardson's stationary model of wars is correct in an absolute sense. Considerable evidence indicates that dyads do not independently generate conflicts, and that dyadic likelihoods vary across time and space, and by national covariates (Ray, 1998; Ward et al., 2007; Leeds, 2003; Alesina & Spolaore, 1997; Jackson & Nei, 2015).

Similarly, normalizing a war's size by the nations' or world's total population, producing a per-capita rate (Pinker, 2012; Cirillo & Taleb, 2015), assumes that individuals contribute independently and with equal probability to potential or actual violence, regardless of who or where they are. Normalizing by world population is thus equivalent to assuming that doubling Canada's population would linearly increase the level of violence in a war in Yemen. In general, human populations have increased so dramatically over the past 200 years that this normalizer nearly always produces the appearance of a decline in violence, even if wars were, in an absolute sense, stationary. However, there is little evidence that real conflict sizes or rates increase linearly with population (Bowles, 2009; Oka et al., 2017; Falk & Hildebolt, 2017).

That said, per-capita variables can be useful for other reasons (Pinker, 2012), and population surely does play some role in the sizes of wars, albeit probably not a simple one (Oka et al., 2017; Falk & Hildebolt, 2017). A realistic per-capita normalizer should instead account for the effects of alliances, geographic proximity, geopolitical stability, democratic governance, economic ties, etc. (Cederman et al., 2011), in addition to population, and would be akin to modeling the underlying processes that generate interstate wars. This represents an important direction for future work.

10.3   The Size and Timing of Wars

In Richardson's view, the size or severity of an interstate war (number of battle deaths) follows a power-law distribution of the form $\Pr(x) \propto x^{-\alpha}$ for some $\alpha > 1$ and for all $x \geq x_{\min} > 0$. Using standard techniques to estimate α and $x_{\min}$, and to test the fitted distribution (Clauset et al., 2009), I find that the set of observed interstate wars sizes, from 1823 to 2003, are statistically indistinguishable from an iid draw from a power-law distribution (Fig. 10.2).

Similarly, Richardson posits that the onset of a new interstate war follows a Poisson process, meaning that wars arrive at a constant rate q, and the time t between onsets of new wars follows a geometric distribution of the form $\Pr(t) \propto e^{-qt}$ for some $q > 0$ and for all $t \geq 1$.

Using the same techniques as above to estimate q and test the fitted distribution, I find that the set of observed delays between onsets are statistically indistinguishable from an iid draw from a Poisson process (Fig. 10.3). That is, both the size and timing of interstate wars 1823–2003 are statistically indistinguishable from

118 A. Clauset

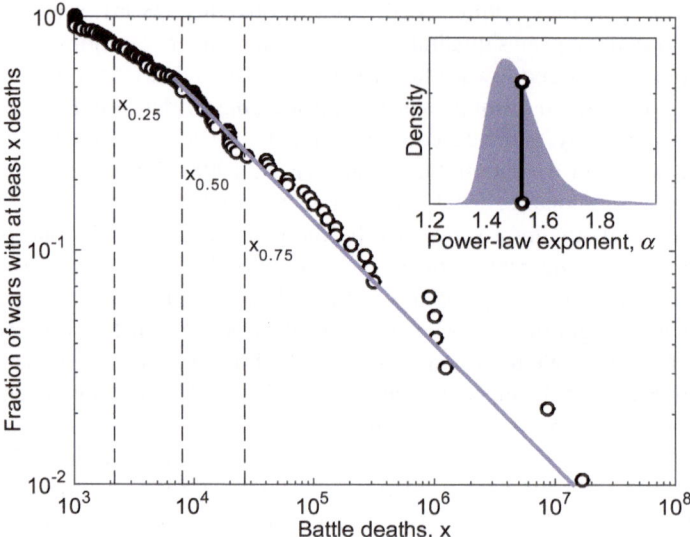

Fig. 10.2 Interstate wars sizes, 1823–2003. The maximum likelihood power-law model of the largest-severity wars (solid line, $\alpha = 1.53 \pm 0.07$ for $x_{min} = 7061$) is a statistically plausible data-generating process of these 51 empirical severities (Monte Carlo, $p_{KS} = 0.78 \pm 0.03$). For reference, distribution quartiles are marked by vertical dashed lines. Inset: bootstrap distribution of maximum likelihood parameters $\Pr(\alpha)$, with the empirical value (black line). Originally published in Clauset (2018)

Richardson's simple model of a power-law distribution for war sizes and a Poisson process for their arrival (Richardson, 1944, 1960). This agreement is remarkable considering the overall simplicity of the model compared to the complexity and contingency of international relations over this period, and the fact that this time period includes nearly 60 years of additional data over Richardson's original analysis.

10.4 Are Large Wars Declining?

Combining the two parts of Richardson's model allows us to generate simulated interstate war data sets, drawn from a stationary process, with similar onset times and war sizes as the historical record. The statistics of these simulated histories define a reference distribution against which we can compare aspects of the historical record.

I now apply this model to address a long-running debate in international relations: did the underlying processes that generate interstate wars change after the

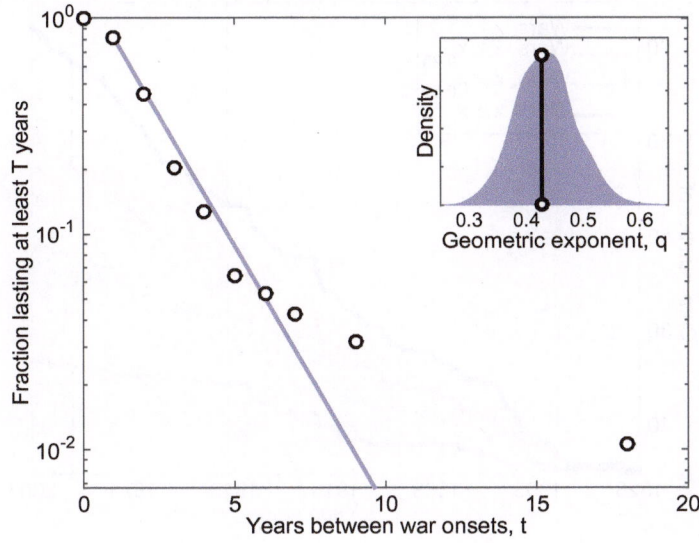

Fig. 10.3 Times between interstate war onsets, 1823–2003. The maximum likelihood geometric model (solid line, $q = 0.428 \pm 0.002$ for $t \geq 1$) is a statistically plausible data-generating process of the empirical delays (Monte Carlo, $p_{KS} = 0.13 \pm 0.01$), implying that the apparent discontinuity at $t = 5$ is a statistical artifact. Inset: the bootstrap distribution of maximum likelihood parameters $\Pr(q)$, with the empirical estimate (black line). Originally published in Clauset (2018)

Second World War? This point in history is commonly proposed as the beginning of a 'long peace' pattern in interstate conflict, meaning a pronounced decrease in the frequency and severity of wars, especially large ones (Gaddis, 1986; Ward et al., 2007; Levy & Thompson, 2010; Pinker, 2012; Braumoeller, 2013).

To test the long peace hypothesis from Richardson's perspective, I consider the accumulation of large interstate wars over time, and assess whether or when that accumulation in the post-war period represents a low probability event under the stationary model. If the accumulation during the long peace is statistically unusual under the reference distribution, it would indicate support for an underlying change in the generating processes.

I define 'large' wars as those in the upper quartile of the historical war size distribution, meaning $x \geq x_{0.75} = 26{,}625$ battle deaths, but similar results are obtained for other large thresholds. The initial 1823–1939 period contains 19 such large wars, for an arrival rate of one per 6.2 years, on average. The 'great violence' pattern of 1914–39, which spans the onsets of the First and Second World Wars, includes ten large wars, or about one every 2.7 years. The long peace of the 1940–2003 post-war period contains only five large wars, or about one every 12.8 years. Figure 10.4 shows the historical accumulation curves of these events, and for smaller wars, as a function of time.

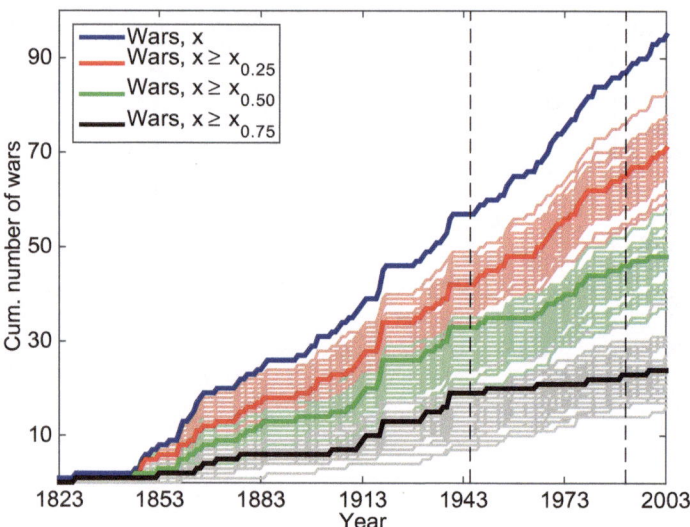

Fig. 10.4 Historical and simulated accumulation curves of interstate wars. Empirical cumulative counts of wars of different sizes (dark lines) over time, alongside ensembles of simulated counts from a stationary model (light lines), in which empirical severities are replaced iid with a bootstrap draw from the empirical severity distribution. Dashed lines mark the end of the Second World War and the end of the Cold War. Originally published in Clauset (2018)

Our combined model takes the historical war onset years as given, and then for each of these 95 conflicts, draws a synthetic war size iid from the empirical size distribution (with replacement), as in a simple bootstrap of the data. Clauset (2018) considers two additional models of this flavor, which produce larger variances in the accumulation curves for large wars but yield similar results and conclusions.

10.5   Evaluating the Past

Within the historical accumulation curve for large wars, the long peace is a visible pattern, in which the arrival rate (the curve's slope) is substantially flatter than in the preceding great violence period (Fig. 10.4). However, under the stationary model, this pattern is well within the envelope of simulated curves, and the observed pattern is statistically indistinguishable from a typical excursion, given the heavy-tailed nature of historical war sizes.

In fact, most simulated war sequences contain a period of 'peace' at least as long in years and at least a peaceful in large-war counts as the long peace (Table 10.1). Fifty years or more of relatively few large wars is thus entirely typical, given the empirical distribution of war sizes, and observing a long period of peace is not necessarily evidence of a changing likelihood for large wars (Cirillo & Taleb, 2015;

Table 10.1 Stationary likelihood of empirical conflict patterns. Under a simple stationary model of conflict generation (see text), the estimated likelihoods of observing two particular large-war patterns over the period 1823–2003: a great violence, meaning 10 or more large war onsets ($x \geq x_{0.75}$) over a 27 year period (the empirical count of such onsets, 1914–39); or, a long peace, meaning 5 or fewer large war onsets over a 64 year period (the empirical count of such onsets, 1940–2003). Probabilities estimated by Monte Carlo. Parenthetical values indicate the standard error of the least significant digit. Originally published in Clauset (2018)

Empirical pattern	Formalization	Model
Great violence	$\Pr(V \equiv n \geq 10$ large wars over $t \leq 27$ years)	0.107(1)
Long peace	$\Pr(P \equiv n \leq 5$ large wars over $t \geq 64$ years)	0.622(2)

Clauset, 2018). Even periods comparable to the great violence of the World Wars are not statistically rare under Richardson's model (Table 10.1).

Taking these findings at face value implies that the probability of a very large war is constant. Under the model, the 100-year probability of at least one war with 16,634,907 or more battle deaths (the size of the Second World War) is 0.43 ± 0.01, implying about one such war per 161 years, on average.

10.6   Peering into the Future

We can also use Richardson's model to simulate future sequences of interstate wars, and thereby evaluate how long the long peace must last before it becomes compelling evidence that the underlying processes did, indeed, change after the Second World War. To extend the simulated war sequences beyond 2003, for each year, I create a new war onset according to a simple Bernoulli process, with the historical production rate (on average, a new war every 1.91 years). I then linearly extrapolate the long peace pattern, in which a new large war occurs on average every 12.8 years, out into the future until 95% of the simulated accumulation curves exceed the extrapolated pattern's curve. At that moment in time, the long peace pattern will have become statistically significant, by conventional standards, relative to the stationary model, and we could say with confidence that the time since the Second World War was governed by a different, more peaceful underlying process.

In this extrapolated future, the post-war pattern of relatively few large wars becomes progressively more unlikely under a stationary hypothesis (Fig. 10.5). However, it is not until 100 years into the future that the long peace becomes statistically distinguishable from a large but random fluctuation in an otherwise stationary process. Even if there were no large wars anywhere in the world after 2003, the year of significance would arrive only a few decades earlier.

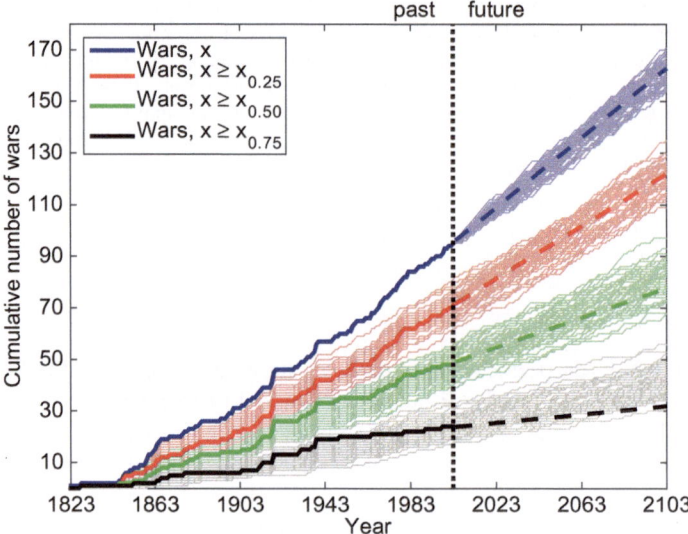

Fig. 10.5 How long must the peace last? Simulated accumulation curves for wars of different sizes under a simple stationary model, overlaid by the empirical curves up to 2003 (dark lines) and linear extrapolations of the empirical post-war trends (the long peace) for the next 100 years (dashed lines). *Source* Originally published in Clauset (2018)

The consistency of the historical record of interstate wars with Richardson's stationary model places an implicit upper bound on the magnitude of change in the underlying conflict generating process since the end of the Second World War (Cederman, 2001). This modeling effort cannot rule out the existence of a change in the rules that generate interstate conflicts, but if it occurred, it cannot have been a dramatic shift. The results here are entirely consistent with other evidence of genuine changes in the international system, but they constrain the extent to which such changes could have genuinely impacted the global production of interstate wars.

10.7   Discussion

The agreement between the historical record of interstate wars and Richardson's simple model of their frequency and severity is truly remarkable, and it stands as a testament to Richardson's lasting contribution to the study of violent political conflict.

There are, however, a number of caveats, insights, and questions that come out of our analysis. For instance, Richardson's Law – a power-law distribution in conflict event sizes – appears to hold only for sufficiently large 'deadly quarrels,' specifically those with 7061 or more battle deaths. The lower portion of the distribution is slightly more curved than expected for a simple power law, which suggests potential differences in the processes that generate wars above and below this threshold.

With only 95 conflicts and a heavy-tailed distribution of war sizes, there are relatively few large wars to consider. This modest sample size surely lowers the statistical power of any test and is likely partly to blame for needing nearly 100 more years to know whether the long peace pattern is more than a run of good luck under a stationary process.

One could imagine increasing the sample size by including civil wars, which are about three times more numerous than interstate wars over 1823–2003. Including these, however, would confound the resulting interpretation, because civil wars have different underlying causes (Salehyan & Gleditsch, 2006; Cederman et al., 2013; Wucherpfennig et al., 2016), and because the distribution of civil war sizes is shifted toward smaller conflicts and exhibits relatively fewer large ones (Lacina, 2006).

Putting aside these technical issues, the larger question our analysis presents is this: how can it be possible that the frequency and severity of interstate wars are so consistent with a stationary model, despite the enormous changes and obviously non-stationary dynamics in human population, in the number of recognized states, in commerce, communication, public health, and technology, and even in the modes of war itself? The fact that the absolute number and sizes of wars are plausibly stable in the face of these changes is a profound mystery for which we have no explanation.

There is, of course, substantial evidence for a genuine post-war trend toward peace, based on mechanisms that reduce the likelihood of war (Ray, 1998; Leeds, 2003; Jackson & Nei, 2015) and on statistical signatures of a broad and centuries-long decline in general violence (Gurr, 2000; Payne, 2004; Goldstein, 2011; Pinker, 2012) or the improvement of other aspects of human welfare (Roser et al., 2017).

But a full accounting of the likelihood that the long peace will endure must also consider mechanisms that increase the likelihood of war (for example, see Bremer, 1992; Mansfield & Snyder, 1995; Barbieri, 1996). War-promoting mechanisms certainly include the reverse of established peace-promoting mechanisms, e.g., the unraveling of alliances, the slide of democracies into autocracy, and the fraying of economic ties, but they may also include unknown mechanisms.

In the long run, processes that promote interstate war may be consequences of those that reduce it over the shorter term, through feedback loops, tradeoffs, or backlash effects. For example, the persistent appeal of nationalism, whose spread can increase the risk of interstate war (Schrock-Jacobson, 2012), is not independent of deepening economic ties via globalization (Smith, 1992). Investigating such interactions is a vital direction of future research and will facilitate a more complete understanding of the processes that govern the likelihood of patterns like the long peace.

More concretely, our results here indicate that the post-war efforts to reduce the likelihood of large interstate wars have not yet changed the observed statistics enough to tell if they are working. This fact does not lessen the face-value achievement of the long peace, as a large war today between major powers could be very large indeed, and there are real benefits beyond lives saved (Roser et al., 2017) that have come from increased economic ties, peace-time alliances, and the spread of democracy. However, it does highlight the continued relevance of Richardson's foundational ideas as the appropriate null hypothesis for patterns in interstate war.

One explanation for the apparent stationarity of wars since 1823 is the existence of compensatory trends in related conflict variables that mask a genuine change in the conflict generating processes. Patterns across multiple conflict variables do seem to indicate a broad shift toward less violence (Gurr, 2000; Payne, 2004; Goldstein, 2011; Pinker, 2012). But, not all conflict variables support this conclusion, and some, such as military disputes and the frequency of terrorism, seem to be increasing, instead (Harrison & Wolf, 2012; Clauset & Woodard, 2013). Untangling conflict variables' interactions and characterizing their trends and differences across groups of nations will be a valuable line of future work.

An alternative explanation is that the mechanisms that govern the likelihood of war have unfolded heterogeneously across time and geographic regions over the past 200 years, thereby creating an illusion of global stationary by coincidence. The long peace pattern is sometimes described only in terms of peace among largely European powers, who fell into a peaceful configuration after the great violence for well understood reasons. In parallel, however, conflicts in other parts of the world, most notably Africa, the Middle East, and Southeast Asia, have become more common, and these may have statistically balanced the books globally against the decrease in frequency in the West, and may even be causally dependent on the drivers of European war and then peace.

Developing a more mechanistic understanding of how changes in the likelihood of conflict in one part of the world may induce compensatory changes in the likelihood of conflict in other parts would have enormous value if it could explain how some regions can fall into more peaceful patterns as a group, while other regions go the opposite direction. The evident stability of Richardson's Law may, in the end, be an artifact of these kinds of complex, 'macro' scale dynamics, playing out across the global stage.

Finally, it is worth reiterating how remarkable and counter-intuitive it is that Richardson's original models of the frequency and severity of wars, first proposed more than half a century ago, successfully hold up under modern, more rigorous statistical methods of evaluation applied to far more comprehensive data. A sobering implication of this success is that the probability of a large interstate appears to have remained constant, despite profound collective efforts to lower it.

More importantly for the general study of conflict, Richardson's work presents a simple and enduring mystery: how can the frequency and severity of interstate wars be so consistent with a stationary model, despite the dramatic changes and obviously non-stationary dynamics in so many other aspects of human civilization? Answering this question will shed new light on the underlying causes of war, and greatly inform efforts to promote peace. Richardson, who was an avowed pacifist and who worked as an ambulance driver during the First World War, would surely be pleased if his work on the statistics of war ultimately helped devise better policies to promote peace.

References

Alesina, Alberto & Enrico Spolaore (1997) On the number and size of nations. *Quarterly Journal of Economics* 112(4): 1027–1056.

Barbieri, Katherine (1996) Economic interdependence: A path to peace or a source of interstate conflict? *Journal of Peace Research* 33(1): 29–49.

Biggs, Michael (2005) Strikes as forest fires: Chicago and Paris in the late 19th century. *American Journal of Sociology* 111(1): 1684–1714.

Bohorquez, Juan C; Sean Gourley, Alexander R Dixon, Michael Spagat & Neil F Johnson (2009) Common ecology quantifies human insurgency. *Nature* 462(7275): 911–914.

Bowles, Samuel (2009) Did warfare among ancestral hunter-gatherers affect the evolution of human social behaviors? *Science* 324(5932): 1293–1298.

Braumoeller, Bear F (2013) Is war disappearing? APSA Chicago 2013 Meeting. Available at SSRN: https://ssrn.com/abstract=2317269.

Bremer, Stuart A (1992) Dangerous dyads: Conditions affecting the likelihood of interstate war, 1816–1965. *Journal of Conflict Research* 36(2): 309–341.

Cederman, Lars-Erik (2001) Back to Kant: Reinterpreting the democratic peace as a macro-historical learning process. *American Political Science Review* 95(1): 15–31.

Cederman, Lars-Erik (2003) Modeling the size of wars: From billiard balls to sandpiles. *American Political Science Review* 97(1): 135–150.

Cederman, Lars-Erik; Kristian Skrede Gleditsch & Halvard Buhaug (2013) *Inequality, Grievances, and Civil War*. Cambridge: Cambridge University Press.

Cederman, Lars-Erik; T Camber Warren & Didier Sornette (2011) Testing Clausewitz: Nationalism, mass mobilization, and the severity of war. *International Organization* 65(4): 605–638.

Cirillo, Pasquale & Nassim Nicholas Taleb (2015) On the statistical properties and tail risk of violent conflicts. *Physica A* 429: 252–260.

Clauset, Aaron (2018) Trends and fluctuations in the severity of interstate wars. *Science Advances* 4: eaao3580.

Clauset, Aaron; Cosma R Shalizi, & M E J Newman (2009) Power-law distributions in empirical data. *SIAM Review* 51(4): 661–703.

Clauset, Aaron & Ryan Woodard (2013) Estimating the historical and future probabilities of large terrorist events. *Annals of Applied Statistics* 7(4): 1838–1865.

Clauset, Aaron, Maxwell Young & Kristian Skrede Gleditsch (2007) On the frequency of severe terrorist events. *Journal of Conflict Resolution* 51(1): 58–87.

Falk, Dean & Charles Hildebolt (2017) Annual war deaths in small-scale versus state societies scale with population size rather than violence. *Current Anthropology* 58(6): 805–813.

Gaddis, John Lewis (1986) The long peace: Elements of stability in the postwar international system. *International Security* 10(4): 99–14.

Gleditsch, Kristian Skrede & Aaron Clauset (2018) Trends in conflict: What do we know and what can we know? In: WC Wohlforth & A Gheciu (eds) *Handbook of International Security*. Oxford: Oxford University Press, 227–244.

Goldstein, Joshua S (2011) *Winning the War on War*. New York: Dutton.

Gurr, Ted Robert (2000) Ethnic warfare on the wane. *Foreign Affairs* 79(3): 52–64.

Harrison, Mark & Nikolaus Wolf (2012) The frequency of wars. *Economic History Review* 65(3): 1055–1076.

Jackson, Matthew O & Stephen M Nei (2015) Networks of military alliances, wars, and international trade. *PNAS* 112(50): 15277–15284.

Lacina, Bethany (2006) Explaining the severity of civil wars. *Journal of Conflict Resolution* 50(2): 276–289.

Leeds, Brett Ashley (2003) Do alliances deter aggression? The influence of military alliances on the initiation of militarized interstate disputes. *American Journal of Political Science* 47(3): 427–439.

Levy, Jack S & William R Thompson (2010) *Causes of War*. New York: Wiley-Blackwell.

Lorenz, Edward Norton (1963) Deterministic nonperiodic flow. *Journal of the Atmospheric Sciences* 20(2): 130–141.

Mandelbrot, Benoit (1967) How long is the coast of Britain? Statistical self-similarity and fractional dimension. *Science* 156: 3775.

Mansfield, Edward D & Jack Snyder (1995) Democratization and the danger of war. *International Security* 20(1): 5–38.

Mitchell, Melanie (2011) *Complexity: A Guided Tour*. Oxford University Press.

Newman, MEJ (2005) Power laws, Pareto distributions and Zipf's law. *Contemporary Physics* 46 (5): 323–351.

Oka, Rahul; Marc Kissel, Mark Golitko, Susan Guise Sheridan, Nam C Kim & Agustín Fuentes (2017) Population is the main driver of war group size and conflict casualties. *Proceedings of the National Academy of Science USA* 114(52): E11101–E11110.

Payne, James L (2004) *A History of Force: Exploring the Worldwide Movement against Habits of Coercion, Bloodshed, and Mayhem*. Sandpoint, ID: Lytton.

Pinker, Steven (2012) *The Better Angels of Our Nature*. New York: Penguin.

Ray, James Lee (1998) Does democracy cause peace? *Annual Review of Political Science* 1: 27–46.

Reed, William J & Barry D Hughes (2002) From gene families and genera to incomes and internet file sizes: Why power laws are so common in nature. *Physical Review E* 66(6): 067103.

Resnick, Sidney I (2006) *Heavy-Tail Phenomena: Probabilistic and Statistical Modeling*. New York: Springer.

Richardson, Lewis Fry (1922) *Weather Prediction by Numerical Process*. Cambridge: Cambridge University Press.

Richardson, Lewis Fry (1944) The distribution of wars in time. *Journal of the Royal Statistical Society* 57: 242–250.

Richardson, Lewis Fry (1948) Variation of the frequency of fatal quarrels with magnitude. *Journal of the American Statistical Association* 43(244): 523–546.

Richardson, Lewis Fry (1960) *Statistics of Deadly Quarrels*. Pittsburgh, PA: Boxwood.

Roser, Max; Esteban Ortiz Ospina & Jaiden Mispy (2017) Our World in Data, https://ourworldindata.org.

Salehyan, Idean & Kristian Skrede Gleditsch (2006) Refugees and the spread of civil war. *International Organization* 60(2): 335–366.

Sarkees, Meredith Reid & Frank Wayman (2010) *Resort to War: 1816–2007*. Washington, DC: CQ Press.

Schrock-Jacobson, Gretchen (2012) The violent consequences of the nation: Nationalism and the initiation of interstate war. *Journal of Conflict Resolution* 56(5): 825–852.

Smith, Anthony D (1992) National identity and the idea of European unity. *International Affairs* 68(1): 55–76.

Ward, Michael D; Randolph M Siverson & Xun Cao (2007) Disputes, democracies, and dependencies: A reexamination of the Kantian peace. *American Journal of Political Science* 51 (3): 583–601.

Wucherpfennig, Julian, Philipp Hunziker & Lars-Erik Cederman (2016) Who inherits the state? Colonial rule and postcolonial conflict. *American Journal of Political Science* 60(4): 882–898.

Aaron Clauset, b. 1979, Ph.D. in Computer Science (University of New Mexico, 2006); Omidyar Fellow, Santa Fe Institute (2006–10); Associate Professor of Computer Science and Core Faculty in the BioFrontiers Institute, University of Colorado Boulder (2010–); External Faculty, Santa Fe Institute (2012–), aaron.clauset@colorado.edu.

Chapter 11
The Decline of War Since 1950: New Evidence

Michael Spagat and Stijn van Weezel

Abstract For the past 70 years, there has been a downward trend in war sizes, but the idea of an enduring 'long peace' remains controversial. Some recent contributions suggest that observed war patterns, including the long peace, could have resulted from a long-standing and unchanging war-generating process, an idea rooted in Lewis F Richardson's pioneering work on war. Focusing on the hypothesis that the war sizes after the Second World War are generated by the same mechanism that generated war sizes before the Second World War, recent work failed to reject this 'no-change' hypothesis. In this chapter, we transform the war-size data into units of battle deaths per 100,000 of world population rather than absolute battle deaths – units appropriate for investigating the probability that a random person will die in a war. This change tilts the evidence towards rejecting no-change hypotheses. We also show that sliding the candidate break point slightly forward in time, to 1950 rather than 1945, leads us further down the path toward formal rejection of a large number of no-change hypotheses. We expand the range of wars considered to include not just inter-state wars, as is commonly done, but also intra-state wars. Now we do formally reject many versions of the no-change hypothesis. Finally, we show that our results do not depend on the choice of war dataset.

11.1 A Continuing Debate

The possibility that war might be in decline has long tantalized academics and the general public. Ongoing debate has focused on whether there might be a secular downward trend in war sizes which might herald the decline of war. For roughly 70 years there has not been a truly huge war or a direct confrontation between major powers. Nevertheless, the idea of an enduring 'long peace', in the coinage of Gaddis (1986), remains controversial. Some scholars have developed a decline-of-war thesis in

© The Author(s) 2020   129
N. P. Gleditsch (ed.), *Lewis Fry Richardson: His Intellectual Legacy
and Influence in the Social Sciences*, Pioneers in Arts, Humanities, Science,
Engineering, Practice 27, https://doi.org/10.1007/978-3-030-31589-4_11

some detail (Goldstein, 2011; Pinker, 2011; Hathaway & Shapiro, 2017) while others reject it (Braumoeller, 2013; Cirillo & Taleb, 2016b; Clauset, 2018, 2020). Here we do not attempt a broad survey of the existing literature. Rather, we focus on the recent contributions of Cirillo & Taleb (2016b) & Clauset (2018, 2020 in this volume) suggesting that observed war patterns, including the long peace, could have come from a long-standing and unchanging war-generating process. In particular, we engage with Clauset (2018) who tests the hypothesis that the war sizes after the Second World War are generated by the same mechanism that generated war sizes before the Second World War. He fails to reject what we will call a 'no-change hypothesis'.

Here are the main contributions of our chapter. First, we give a simple exposition of the central ideas behind the new critiques of the decline-of-war thesis made by Cirillo & Taleb (2016b) and Clauset (2018, 2020). These ideas hinge centrally on the original insight of Richardson (1948, 1960) into the fat-tailed size distribution of modern wars. Second, we transform the war-size data into units of battle deaths per 100,000 of world population rather than absolute battle deaths and argue that these units are appropriate for investigating the probability that a random person will die in a war. We show that this change tilts the evidence towards rejecting a large number of no-change hypotheses. Third, we show that sliding the candidate break point slightly forward in time, to 1950 rather than 1945, leads us further down the path toward formal rejection of a range of no-change hypotheses. Finally, we expand the types of wars to include intra-state as well as inter-state. Now we almost always formally reject our no-change hypotheses.[1] Finally, we show that our results do not depend on the choice between two widely used war datasets.

11.2   Richardson Provides Our Framework

Decades ago, Richardson (1948) introduced the idea that war sizes tend to follow what is known as a power law distribution.[2] Technically, this means that the frequency of wars of size x is proportional to $x^{-\alpha}$ where $\alpha > 1$ is some constant. Thus, bigger wars are less common than smaller ones with the value of α governing the rate at which war frequencies decrease as war sizes increase. This remarkable insight has fared well against more than half a century of new data and the development of more rigorous statistical methods for estimating and testing power laws (Cederman, 2003; Clauset, 2018; González-Val, 2016).

For our purpose, the important characteristic of power-law distributions is that they have what are known as 'fat upper tails' governing the relationship between

---

[1]Our findings do not refute those of Clauset (2018). It can be true simultaneously that per capita war sizes decrease while the absolute war size generation mechanism does not change.

[2]Spagat (2015) provides a non-technical introduction to power laws.

war sizes and their frequencies. This property entails that, although bigger wars are less common than smaller ones, the rate at which war frequencies decline with war sizes is much slower than would be the case if war sizes followed a common normal, or 'Bell Curve', distribution. Most people are conditioned to think in terms of Bell Curves, so some mental effort is required to adjust to fat tails. Here is the most salient point to bear in mind in the present context; huge wars are really rare but not really really really, rare.

We illustrate the key fat-tail property with the following numerical example. Suppose that every time the world experiences a new war, w, the probability that the war size will grow to at least the size of the First World War, $\bar{w}$ – hereafter a 'truly huge war' – is 0.006.[3] We now make the important assumption that war-size realizations are statistically independent of each other, which implies that the size of war w tells us nothing about the sizes of previous or future wars. Under these conditions, the chance that there is at least one truly huge war after 200 war-size realizations is roughly 2/3.[4] If we lower the probability that each new war will turn out to be a truly huge one down from 0.006 down to $P(w \geq \bar{w}) = 0.0001$, then the chance of at least one truly huge war in 200 draws drops to around 1 in 50. Decreasing the probability of a truly huge war on each draw even further down to $P(w \geq \bar{w}) = 10^{-7}$, decreases this chance all the way down to about 1 in 50,000. Thus, it makes a big practical difference whether truly huge wars are really rare, $P(w \geq \bar{w}) = 0.006$; really really rare, $P(w \geq \bar{w}) = 0.0001$; or really really really rare, $P(w \geq \bar{w}) = 10^{-7}$.

This fat-tail property of the war-size distribution potentially places the world into what we might call a 'bad Goldilocks' range. On the one hand, 0.006 is large enough that we might expect to suffer a truly huge war once every few generations, far too often for such a calamity. On the other hand, 0.006 is small enough that the risk of a truly huge war can lurk below the surface for a long time without being exposed as a major threat. This is evident within our example according to which the world has about a 1/3 chance of experiencing 200 wars without suffering a truly huge one. And if our luck holds out this long, we could easily last another 200 wars without suffering a truly huge war.

Thus, we arrive at an important insight flowing from the pioneering work of Richardson (1948) and developed further by Clauset (2018); the threat of a truly huge future war can be quite serious while simultaneously remaining well-hidden for a long time. In other words, we should not dismiss the possibility of a truly huge future war just because such an event would be dramatically out of line with our range of experience over the last 70 years. At the same time, we must not imprison ourselves in our own ahistorical assumptions that rely on the artifice of independent

---

[3]This probability is not entirely fictitious. In the dataset compiled by Gleditsch (2004), the First and Second World Wars are by far the biggest two out of 362 wars that occurred between the beginning of the 19th century and 1945: $2/362 \approx 0.006$.

[4]$1 - (1 - 0.006)^{200} \approx 0.7$.

draws with fixed and unchanging probabilities. These calculations are helpful to understand important concepts and establish baseline expectations. But they do not possess any special powers to describe the world we currently live in or to predict its future. A finding that the war-size pattern of recent decades is consistent with an unchanging war generation mechanism over the last two centuries does not prove that that such a mechanism actually exists.

11.3   A New Debate on the Decline of War

There is diversity of opinion among proponents of the decline-of-war thesis. First, it is standard to claim that the absolute level of war violence has declined over time, albeit unevenly (Lacina & Gleditsch, 2005; Human Security Report Project, 2011). Different scholars emphasize different time periods, although most view the Second World War as an important turning point. Second, sometimes the main claim is about per capita, rather than total, war violence (Pinker, 2011). A third tendency is that no one we are aware of argues that truly huge wars have become impossible. To be sure, a sense of optimism pervades this literature with proponents generally providing reasons why war violence is decreasing and why this trend might reasonably be expected to continue. Yet, invariably, there is also a note of caution about the future.

The recent critique of the decline-of-war thesis was instigated by Cirillo & Taleb (2016b), who collected data on 565 wars going all the way back to Boudicca's rebellion against the Romans in the first century common era (CE). Using extreme value theory to fit the fat-tailed data, they find that they cannot reject their model and conclude from this non-rejection that the data do not support a decline-of-war thesis. In a companion paper they go further, writing that 'there is no scientific basis for narratives about change in risk' (Cirillo & Taleb, 2016a).

Cirillo & Taleb (2016b) helped to prompt renewed focus on the importance of fat tails in war sizes for the decline-of-war debate; however, they left several important issues unresolved. First, although a main contribution of their work is the data collection effort, their dataset is not publicly available, and they have refused to allow other researchers to examine it (Spagat, 2017). This stance takes their work outside the scientific universe, at least for now. Second, non-rejection of a model fitting two thousand years of data does not rule out the possibility of scientifically grounded discussions about possible changes in war risks during subsets of these two thousand years. For example, there could be a big change after, e.g., war number 500 but without the last 65 draws disturbing the fit of the first 500 draws sufficiently to lead to rejection of the whole model. Imagine flipping a coin that has a 0.5 probability of landing heads for the first 500 flips and a 0.3 probability of landing heads for the last 65. You would probably not reject a hypothesis that all the flips had a chance pretty close to 0.5 of landing heads. More importantly, if you confine your analysis to the 565 flips as a whole, then you will get no hint that there was a dramatic change after flip number 500. It would have been more appropriate to test for a break in the data at a potential change point, such as the end of the Second World War; Cirillo & Taleb (2016b) do not provide such a test. Third, there is an overarching assumption in this approach that the only evidence scientifically admissible to our discussion is a list of war sizes and timings. Cirillo & Taleb (2016b) seem to think that historical events such as peace treaties, formation of international institutions or social trends such as improving human rights are, simply, outside the bounds of a scientific discussion; this restrictive view makes little sense.

Clauset (2018) addresses the first two of the unresolved issues. First, he uses the open-source Correlates of War (COW) dataset that covers interstate wars from the beginning of the 19th century to the present (2007). Second, his whole analysis focuses on testing for a trend break starting at the end of the Second World War. The essence of his approach on war sizes is to fit a power law to the data up through the Second World War and then test the hypothesis that the data after 1945 was generated by this distribution, i.e., he tests what we call a no-change hypothesis. Clauset (2018) concludes that he cannot reject the no-change hypothesis. This finding is intuitive in light the numerical examples provided above although there is certainly tension between the no-change hypothesis and the last 70 years.

Clauset (2018) provides a useful contribution to our thinking but, at the same time, we must be cautious about this result for several reasons. First, other information besides the time series of war sizes is potentially relevant. Second, we should not think exclusively in terms of any one particular hypothesis such as the no-change one. There are other hypotheses, more in line with a decline-of-war thesis, that would also not be rejected by the data. For example, suppose we modify the no-change hypothesis by stipulating that wars with more than 5 million battle deaths became very very rare after the Second World War. That is, we virtually eliminate the fat tail from the hypothesized war generation mechanism. This restriction is fully consistent with the post-1945 experience since no war during this period comes close to such a size. Thus, this hypothesis is consistent with decline-of-war ideas and will also not get rejected by the data. And there is no reason to privilege the no-change hypothesis over this one. Third, we must not fall into the trap of accepting the null hypothesis based on its non-rejection. Clauset (2018) finds that we would finally reject his no-change hypothesis ($p < 0.05$) after about 100–140 more years without a truly huge war. Even then we still would not be able to entirely rule out the no-change hypothesis. However, if the data became extremely contrary to the no-change hypothesis after 100 sufficiently peaceful years then the data would already be fairly contrary to this hypothesis after 50 sufficiently peaceful years. Returning to our earlier calculations, recall that the Gleditsch (2004) dataset contains 212 wars for the period after the Second World War. If a further

212 wars occur without a truly huge one, perhaps over the next 70 years, we could then reject this version of the no-change hypothesis at a 10% level, which would be rather convincing evidence that there was a change for the better. In other words, the 0.05 threshold is arbitrary and excessively binary; non-rejection of the no-change hypothesis does not mean that the decline-of-war thesis is false until it suddenly switches to true after 100 years without a truly huge war.

11.4   Measuring War

Our empirical analysis relies on two datasets that cover war sizes and dates; the commonly used Correlates of War (COW) dataset (Sarkees & Wayman, 2010), which was also used by Clauset (2018), and the dataset compiled by Gleditsch (2004). The two datasets overlap substantially, and both cover the period 1816–2007. Indeed, the Gleditsch (2004) dataset is based on the COW dataset. However, there are important distinctions that are worth understanding even though it turns out that our results do not depend materially on the choice of dataset. For COW there is a big change in the inclusion criteria in 1920 with the founding of the League of Nations. The fundamental test for COW is always membership in the international system for both states, in the case of inter-state war, and for the state in the case of intra-state war. Between 1816 and 1920 this test breaks down into two parts; (i) a population greater than 500,000 and (ii) being 'sufficiently unencumbered by legal, military, economic, or political constraints to exercise a fair degree of sovereignty and independence'. After 1920, the COW test switches to membership in the League of Nations (or United Nations) and receiving diplomatic missions from any two major powers (Singer & Small, 1972). Gleditsch & Ward (1999) note that, in practice, the pre-1920 test boils down to having formal diplomatic relations with Britain and France. This rule excludes many countries and their wars, including the three Anglo-Afghan wars that took place between 1839 and 1919 and some intrastate wars such as the 1831–45 civil war in Central America.

It would be unfair to label the COW dataset as simply incorrect, yet we believe that its British-French emphasis excludes many wars that are relevant to the decline-of-war debate. The revised data by Gleditsch (2004), which corrects these systematic problems, contains 574 wars between 1816 and 2007, 136 of which are interstate wars. During the same period COW contains only 474 wars, 95 of which are interstate. Thus, the difference in war counts is substantial. Moreover, 1920 is close enough to the Second World War so that the 1920 switch could potentially affect the results of Clauset (2018). Thus, we prefer the Gleditsch data but run our calculations on both datasets.[5]

---

[5]The Gleditsch (2004) dataset covers about two centuries of war yet contains roughly the same number of wars as the Cirillo & Taleb (2016b) dataset which covers two millennia of wars. The inclusion criteria for the two datasets seem to be similar.

Our second departure from Clauset (2018) is that we divide all war sizes by world population estimates. These are applied to the start year of each war and taken from Fink-Jensen (2015), Klein Goldewijk et al. (2010), and UN (2013) with some interpolations before 1950. The probability that an average person will be killed in war is of particular interest to the decline-of-war discussion and population adjustment is appropriate to assess this probability. In a similar vein, analysts normally assess, for instance, violence progress by examining the number of homicides per 100,000 of population or the quality of health services through the number of maternal deaths per 1000 live births. At the same time, we recognize the point of Braumoeller (2013) who argues that examination of unadjusted war sizes is of great relevance to understanding human war-proneness.[6]

A third contrast with Clauset (2018) is that we include in our analysis all the wars in each dataset, not just interstate wars.[7] We think that there is no *a priori* theoretical justification to separate out interstate wars and agree with Small & Singer (1982) who argued that 'an understanding of international war cannot rest on interstate wars alone'. The common focus on wars involving major powers or other interstate wars seems to be driven by data availability rather than theoretical considerations (Cunningham & Lemke, 2013). Indeed, the third, fourth and sixth largest wars measured in per capita terms in the Gleditsch dataset are all intra-state (Fig. 11.1). Thus, combining all wars is best practice in our view although we also run our calculations on interstate wars alone.

War-size numbers are intended to include just battle deaths, but both of our datasets work from available sources that sometimes mix in other kinds of deaths. This issue creates two separate problems. First, ideally we would have data on the full human cost of war but often we only have data on the battle-death component of this cost. For example, both datasets record 910,084 deaths for the Korean War, but a full figure would include famine deaths that could push the number up to 5 or 6 million (Lacina et al., 2006). Second, there is inconsistency across wars since some figures hue close to a battle-deaths-only concept whereas other figures are more comprehensive.

---

[6]A war that kills one million people is an unmitigated disaster both in a world of 5 billion people and in a world of 9 billion people.

[7]For COW, all wars means inter-state, intra-state, extra-systemic and non-state. The Gleditsch dataset does not have the last two of these categories, although its more inclusive definition of state means that it codes some COW extra-systemic and non-state wars as either inter-state or intra-state wars. Arguably, we should subtract the populations of ungoverned spaces that fall outside the scope of the Gleditsch dataset from our world population figures. Such adjustments would enhance our decline-of-war results because they would increase the per capita sizes of earlier wars relative to later ones; governance spreads over time. However, these adjustments would be very hard to perform with any degree of accuracy, so we do not attempt them here.

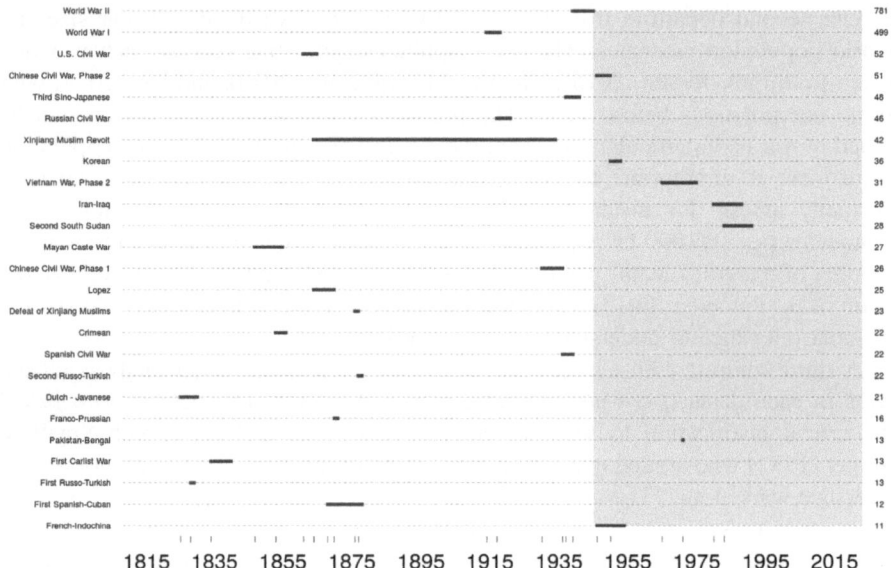

Fig. 11.1 The largest wars as measured by battle deaths per 100,000 on the right y axis. Based on Gleditsch (2004)

11.5 Insights from the Data

A particular feature of our approach is the large number of no-change hypotheses that we test. All our hypotheses are based on two separate cut-off points: one for time periods and the other for per capita war sizes. Our time periods pivot around either the Second World War or the Korean War, but future work should consider more cut-off points. For war sizes we consider all possible cut-offs and examine the fraction of all wars above each war-size cut-off for both the early period and the late period. In short, we examine many right-hand tails and test whether the tails for the later periods are thinner than the tails for the earlier periods.

Here are some sample calculations when the time cut-off point is 1945. According to the Gleditsch (2004) data, there were 362 wars between 1816 and 1945 with the Second World War being by far the largest. Our first no-change hypothesis for the post-1945 period is that the probability that a random war after 1945 will kill at least 781 people per 100,000 (Fig. 11.1) is given by the fraction of all wars before 1945 that reached this violence level. This fraction is $p_0 = \frac{1}{362} \approx 0.003$.

Zero wars out of 212 in the Gleditsch (2004) data attained this size between 1946 and 2007. If war sizes are drawn randomly and independently of each other and if the no-change hypothesis is true, then the probability of this happening is $\left(1 - \frac{1}{362}\right)^{212} = 0.56$. This probability can be interpreted as a p-value on one particular no-change hypothesis at the most extreme end of the distribution of war sizes.[8]

Next we calculate exactly the same types of p-values but for lower and lower war sizes. For war sizes beginning at 781 per 100,000 and moving down towards 499 per 100,000, the size of the First World War, the p-values stay constant. At 499 battle deaths per 100,000 the p-value drops to $(1 - \frac{2}{362})^{212} = 0.31$. It then stays constant all the way down to 52 battle deaths per 100,000, the size of the American Civil War (1861–65), where the p-value drops down to $0.17 - (359/362)^{212}$. In short, the three biggest wars were all before World War II inclusive and together they yield a preponderance of evidence against a no-change hypothesis but not a formal rejection at the 0.05 level. The next largest war is the second phase of the Chinese Civil War which pitted the communists under Mao Zedong against the nationalists under Chiang Kai-shek and caused 51 battle deaths per 100,000 people. The no-change hypothesis assigns probability $\frac{3}{362}$ to the probability that each war size after World War II will exceed 51. This happens once in 212 draws so the p-value on the no-change hypothesis adds together the probability of 0 wars above size 51 and the probability of 1 war above size 51, leading to a p-value of 0.47.

We calculate p-values similarly as we move to smaller and smaller war sizes. When, for example, there are 6 wars before 1945 of size s and above then the no-change hypothesis fixes a probability of $\frac{6}{362}$ on the event that a new post-1945 war will be of size s or above. When, for example, three out of these 212 wars after 1945 are above size s then the p-value on the no-change hypothesis is the probability of three or fewer wars of size s or greater after 212 independent draws, each with probability $\frac{6}{362}$ of reaching this size. We use the binomial formula to make this calculation.[9]

Panel (a) in Fig. 11.2 displays the p-values for the tests of all no-change hypotheses tests with cut-offs for war sizes below 50 battle deaths per 100,000 and with a time break point of 1945. Reading from right to left the curve dips down below 0.2 as we move through the Third Sino-Japanese War which began in 1937,[10] the Russian Civil War (1918–20) following the Russian Revolution of 1917, and the 1864 Muslim revolt in Xinjiang, China. The p-values then rise back above 0.8 because the next four largest wars all occurred after the Second World War. These are the Korean War (1950–53), the second phase of the Vietnam War which started in 1965, the Iran-Iraq war between 1980 and 1988, and finally the Second South Sudan War (1983–2005). Next, continuing to read from right to left, the next 9 wars all took place before the Second World War, bringing the p-values back down to around 0.2.

---

[8]Braumoeller (2013) offers a similar calculation.

[9]For simplicity, we specify our no-change hypotheses as single probabilities rather than as uncertain ranges of probabilities, although we plan to relax this assumption in future research.

[10]This war is often known as the Second Sino-Japanese War. The data counts three wars between China and Japan: the first starting in 1894, the second in 1931, and the third in 1937.

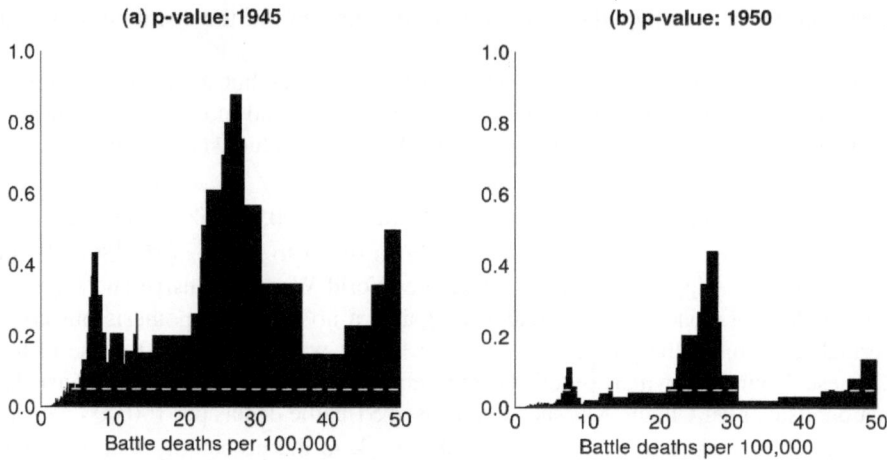

Fig. 11.2 Tests of no-change hypotheses for all wars. Based on Gleditsch (2004), using 1945 (**a**) and 1950 (**b**) as break points

The evidence in Fig. 11.2 is unfavorable to the no-change hypothesis ($p < 0.5$) except in a narrow range of tails for war sizes between about 25 and 28 per 100,000. At the same time, we never reject the no-change hypothesis at the standard 0.05 level. The evidence leans towards the decline of war idea but is far from definitive.

When we use 1950, rather than 1945, as a break point the results are much more favourable to the decline-of-war thesis. Now the eight largest wars in per capita terms all occur before the break point. Panel (b) displays the new p-values. No-change hypotheses are often rejected at 0.05, and even 0.01 levels for a wide range of tails. Two of the very biggest wars (the Chinese Civil War and the Korean War) broke out within the 1945–50 time window so the p-value curve now drops much lower than it did when 1945 was break point.[11]

We have made four separate data changes compared to Clauset (2018): measuring war sizes in per capita terms, using Gleditsch data rather than COW data, considering 1950 as a break point and including intrastate as well as interstate wars. To isolate the importance of each particular change we now consider them in turn. We first note that adjusting for world population levels is essential to get anything resembling the results in this chapter. This is so much true that we do not even bother showing pictures unadjusted for population. Second, the choice of COW or the Gleditsch war data does not matter much (Fig. 11.3). Third, both Figs. 11.2 and 11.3 show that the choice of break point does matter; evidence against the no-change hypothesis is much stronger when the break is at 1950 than it is when the

---

[11]We date wars by when they start.

break point is 1945. Finally, Figs. 11.4 and 11.5 show that our decision to include intrastate wars also matters. We think this is simply due to sample size; excluding intrastate wars decreases the number of wars, making it harder to reject the no-change hypothesis.

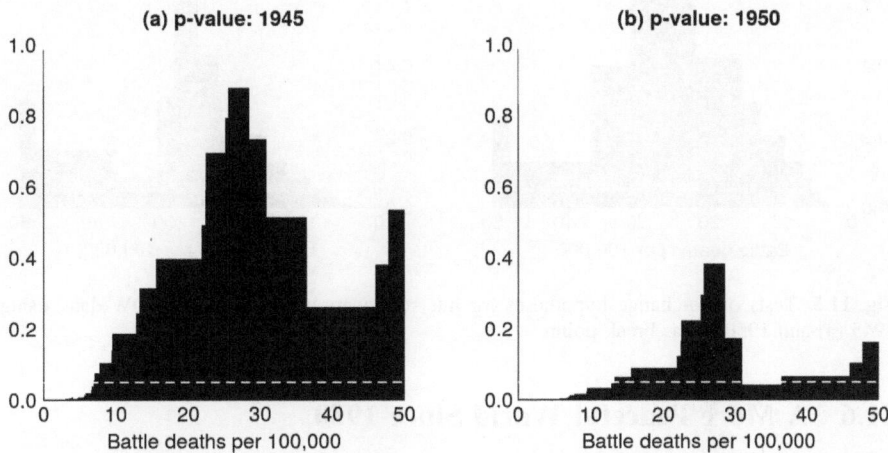

Fig. 11.3 Tests of no-change hypotheses for all wars. Based on COW data, using 1945 (**a**) and 1950 (**b**) as break points

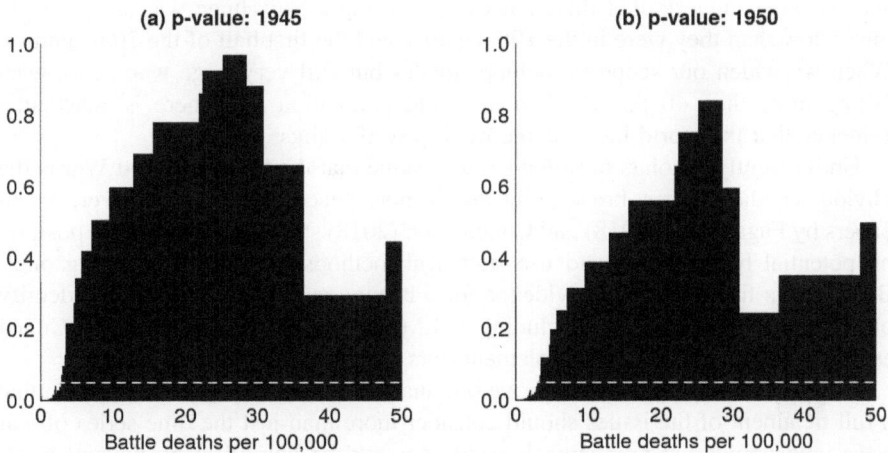

Fig. 11.4 Tests of no-change hypotheses for interstate wars only. Based on Gleditsch (2004), using 1945 (**a**) and 1950 (**b**) as break points

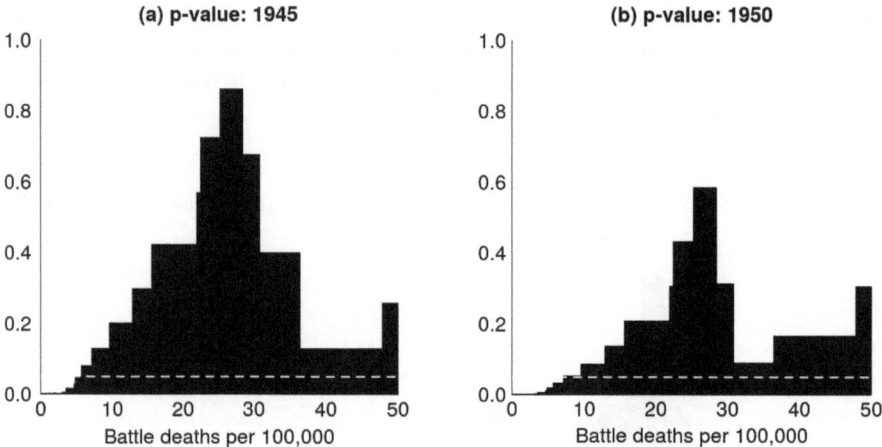

Fig. 11.5 Tests of no-change hypotheses for interstate wars only. Based on COW data, using 1945 (**a**) and 1950 (**b**) as break points

11.6 A More Peaceful World Since 1950

There will continue to be debate on the probability of another truly huge war. If we limit our attention to the probability of a future war at least as large as the First World War then, consistent with Clauset (2018), our analysis suggests that there is presently not enough data to draw a strong conclusion. At the same time, our analysis also suggests that the chances of drawing a truly huge war are probably lower now than they were in the 19th century and the first half of the 20th century. When we widen our scope to include smaller but still very large wars, e.g., wars killing more than 40 per 100,000 of world population then there is substantial evidence that the world has become more peaceful since the 1950s.

Until recently scholars have tended to assume that the Second World War is the obvious candidate for a break point into a more peaceful world. However, recent papers by Fagan et al. (2018) and Cunen et al. (2018) start from an agnostic position on potential break points and use statistical methods to detect convincing ones. Both papers find substantial evidence for a change at 1950 although they identify other candidate break points including 1912 (Fagan et al., 2018) and 1965 (Cunen et al., 2018). These results complement ours nicely.

There is certainly room to improve our analysis. First, we repeat our caution that a full treatment of the issues should consider more than just the time series of war sizes (and population numbers). Second, it would be helpful to go beyond battle deaths to include more complete numbers on war deaths. Unfortunately, it is unlikely that this second hope will ever be fully possible. Third, the new research into defining change points is an important development that will, hopefully, continue. Despite the potential for improvement, we believe that our chapter should shift the debate in favour of the decline-of-war thesis.

References

Braumoeller, Bear (2013) Is war disappearing? Paper presented to APSA Convention, Chicago, IL, available at SSRN https://ssrn.com/abstract=2317269.

Cederman, Lars-Erik (2003) Modeling the size of wars: From billiard balls to sandpiles. *American Political Science Review* 97(1): 135–150.

Cirillo, Pasquale & Nassim Nicholas Taleb (2016a) The decline of violent conflicts: What do the data really say? *Tandon Research Paper* (2876315), available at SSRN https://ssrn.com/abstract=2876315.

Cirillo, Pasquale & Nassim Nicholas Taleb (2016b) On the statistical properties and tail risk of violent conflicts. *Physica A: Statistical Mechanics and Its Applications* 452: 29–45.

Clauset, Aaron (2018) Trends and fluctuations in the severity of interstate wars. *Science Advances* 4: eaao3580.

Clauset, Aaron (2020) On the frequency and severity of interstate wars. Ch. 10 in this volume.

Cunen, Celine; Nils Lid Hjort & Håvard Mokleiv Nygård (2018) Statistical sightings of better angels: Analysing the distribution of battle deaths in interstate conflict over time, www.mn.uio.no/math/english/research/projects/focustat/publications_2/better_angels_finala.pdf.

Cunningham, David E & Douglas Lemke (2013) Combining civil and interstate wars. *International Organization* 67(3): 609–627.

Fagan, Brennen; Marina Knight, Niall MacKay & A Jamie Wood (2018) Change point analysis of historical war deaths. Unpublished paper, Department of Mathematics, University of York.

Fink-Jensen, Jonathan (2015) Total Population. IISH Dataverse. http://hdl.handle.net/10622/SNETZV.

Gaddis, John Lewis (1986) The long peace: Elements of stability in the postwar international system. *International Security* 10(4): 99–142.

Gleditsch, Kristian S (2004) A revised list of wars between and within independent states, 1816–2002. *International Interactions* 30(3): 231–262.

Gleditsch, Kristian S & Michael D Ward (1999) Interstate system membership: A revised list of the independent states since 1816. *International Interactions* 25(4): 393–413.

Goldstein, Joshua (2011) *Winning the War on War: The Decline of Armed Conflict Worldwide.* New York: Penguin.

González-Val, Rafael (2016) War size distribution: Empirical regularities behind conflicts. *Defence and Peace Economics* 27(6): 1–16.

Hathaway, Oona A & Scott J Shapiro (2017) *The Internationalists: How a Radical Plan to Outlaw War Remade the World.* New York: Simon Schuster.

Human Security Report Project (2011) *Human Security Report 2009/2010: The Causes of Peace and the Shrinking Costs of War.* Oxford: Oxford University Press.

Klein Goldewijk, Kees; Arthur Beusen & Peter Janssen (2010) Long term dynamic modeling of global population and built-up area in a spatially explicit way, HYDE 3.1. *Holocene* 20(4): 565–573.

Lacina, Bethany & Nils Petter Gleditsch (2005) Monitoring trends in global combat: A new dataset of battle deaths. *European Journal of Population* 21(2–3): 145–166.

Lacina, Bethany; Nils Petter Gleditsch & Bruce Russett (2006) The declining risk of death in battle. *International Studies Quarterly* 50(3): 673–680.

Pinker, Steven (2011) *The Better Angels of Our Nature: The Decline of Violence in History and Its Causes.* New York: Penguin.

Richardson, Lewis Fry (1948) Variation of the frequency of fatal quarrels with magnitude. *Journal of the American Statistical Association* 43(244): 523–546.

Richardson, Lewis Fry (1960) *Statistics of Deadly Quarrels.* Pittsburgh, PA: Boxwood.

Sarkees, Meredith Reid & Frank Wayman (2010) *Resort to War: 1816–2007.* Washington DC: CQ Press.

Singer, J David & Melvin Small (1972) *The Wages of War, 1816–1965: A Statistical Handbook.* New York: Wiley.

Small, Melvin & J David Singer (1982) *Resort to Arms: International and Civil Wars, 1816–1980.* Beverly Hills, CA: Sage.

Spagat, Michael (2015) Is the Risk of War Declining. May. *Sense about Science USA*, http://senseaboutscienceusa.org/is-the-risk-of-war-declining-2/.

Spagat, Michael (2017) Secret Data Sunday – Nassim Nicholas Taleb Edition, May, https://mikespagat.wordpress.com/2017/05/14/secret-data-sunday-nassim-nicholas-taleb-edition/.

UN (2013) *World Population Prospects: The 2012 Revision.* New York: Population Division, Department of Economic and Social Affairs, United Nations.

Michael Spagat, b. 1960, Ph.D. in Economics (Harvard University, 1988); Professor of Economics, Royal Holloway University of London; most recent article: Fundamental patterns and predictions of event size distributions in modern wars and terrorist campaigns (PLOS ONE, 2018), m.spagat@rhul.ac.uk

Stijn van Weezel, b. 1985, Ph.D. in Economics (University of London, 2015); Postdoctoral research fellow, Radboud university; most recent article: Fundamental patterns and predictions of event size distributions in modern wars and terrorist campaigns (PLOS ONE, 2018), weezel.van@gmail.com

Correction to: Lewis Fry Richardson: His Intellectual Legacy and Influence in the Social Sciences

Nils Petter Gleditsch

Correction to:
N. P. Gleditsch (ed.), *Lewis Fry Richardson: His Intellectual Legacy and Influence in the Social Sciences*, **Pioneers in Arts, Humanities, Science, Engineering, Practice 27, https://doi.org/10.1007/978-3-030-31589-4_8**

In the original version of the book, the following correction were incorporated: The name "Gregory D. Hess" has now been corrected to "George D. Hess" in the pages 70, 83, 98, 111 and 145. The book and the chapters have been updated with the change.

The updated versions of these chapters can be found at
https://doi.org/10.1007/978-3-030-31589-4_6
https://doi.org/10.1007/978-3-030-31589-4_7
https://doi.org/10.1007/978-3-030-31589-4_8
https://doi.org/10.1007/978-3-030-31589-4_9

Richardson's Life and Work

Nils Petter Gleditsch

Richardson's life is described in a number of articles, some by people who knew him personally. Oliver Ashford, a former student, close family friend, and fellow meteorologist, provides the most extensive treatment in a book published some thirty years after Richardson's death (Ashford, 1985). His book also contains a complete bibliography of Richardson's published work as well as a list of ten unpublished papers. Conflict scholars are mostly familiar with Richardson's two posthumously published books (Richardson, 1960a, b). These are also among his most frequently cited works, as noted in Ch. 1. For those who want to study his intellectual development as a peace researcher in greater detail, it will be of interest to follow his writings over a 35-year period, from his brief proposal for a study of mental capacity in adopted children (Richardson, 1913), his examination of voting power in an international assembly (Richardson, 1918) and his first study of the mathematical psychology of war (Richardson, 1919), dedicated to his comrades of the motor ambulance convoy SS Anglaise, 'in whose company this essay was mainly written'. We no longer have to search for the rare original documents, since these and other publications have been reprinted in three massive volumes (Ashford et al., 1993).

> **Box Appendix** A Richardson timeline. Main sources of the timeline: Ashford (1985) and Wikipedia.
>
> 1881, 11 October, Born in Newcastle upon Tyne
> 1893 Bootham School in York, a Quaker boarding school
> 1898 Durham College of Science
> 1900 King's College, University of Cambridge; graduated with a first-class degree
> 1903 Assistant, National Physical Laboratory
> 1905 Junior demonstrator, University College Aberystwyth
> 1906 Chemist, National Peat Industries
> 1907 Meteorologist, National Physical Laboratory

© The Author(s) 2020   143
N. P. Gleditsch (ed.), *Lewis Fry Richardson: His Intellectual Legacy and Influence in the Social Sciences*, Pioneers in Arts, Humanities, Science, Engineering, Practice 27, https://doi.org/10.1007/978-3-030-31589-4

1909 Married Dorothy Garnett (1885–1956)
1909 Head of the chemical and physical laboratory, Sunbeam Lamp Company
1912 Lecturer in Physics, Manchester College of Technology
1913 Superintendent, Eskdalemuir Observatory, Meteorological Office
1916 Ambulance Driver, Friends Ambulance Unit in France
1919 Experimental research, Meteorological Office at Benson, Oxfordshire
1920 Head of the Physics Department, Westminster Training College
1925 BSc in Psychology, University of London, with pure and applied mathematics
1926 DSc in Physics, University of London
1926 Fellow of the Royal Society
1929 MA, University of Cambridge
1929 Special BSc in Psychology, University of London
1929 Principal, Paisley Technical College
1940 Retired in order to devote himself to independent research
1943 Moved to his final home in Kilmun
1953, Summer, Applied for a Research Fellowship at King's College, University of Cambridge
1953, 30 September, Died in Kilmun, Argyll and Bute
1959 Richardson Peace Studies Centre (now Richardson Institute of Peace Studies) founded at Lancaster University[1]
1960 The annual LF Richardson Prize for meritorious papers by young authors in one of the journals of the society established by the Royal Meteorological Society[2]
1997 The Lewis Fry Richardson Medal established by the European Geosciences Union for 'exceptional contributions to nonlinear geophysics in general'[3]
2001 The Lewis F Richardson Lifetime Achievement Award established in order to honor a scholar who has spent most of his/her academic life in Europe and who has made exemplary scholarly contributions to the scientific study of militarized conflict[4]
2015 Lewis F Richardson lecture series inaugurated by Department of Mathematics, University of York[5]

---

[1]www.lancaster.ac.uk/arts-and-social-sciences/research/research-centres/richardson-institute-for-peace-studies/.
[2]www.rmets.org/our-activities/awards/l-f-richardson-prize.
[3]www.egu.eu/awards-medals/lewis-fry-richardson/.
[4]http://ksgleditsch.com/richardson_award.html.
[5]https://www.york.ac.uk/maths/events/lfr/.

Richardson's Main Works

Ashford, Oliver; H Charnock, Julian C R Hunt, Paul Smoker & Ian Sutherland (eds) *Collected Papers of Lewis F Richardson* (1993) Volume 1: *Meteorology and Numerical Analysis* (in two books). Volume 2: In two books with the same title): *Quantitative Psychology and Studies of Conflict*. Cambridge: Cambridge University Press.

Richardson, Lewis F (1922) *Weather Prediction by Numerical Process*. London: Cambridge University Press. Republished 2007 (print), 2010 (electronic). [The original version can be downloaded for free from https://archive.org/details/weatherpredictio00richrich.]

Richardson, Lewis Fry (1960a) *Arms and Insecurity: A Mathematical Study of the Causes and Origins of War*. Edited by Nicolas Rashevsky & Ernest Trucco. Pittsburgh, PA: Boxwood & Chicago, IL: Quadrangle.

Richardson, Lewis Fry (1960b) *Statistics of Deadly Quarrels*. Edited by Quincy Wright & Carl C Lienau. Pittsburgh, PA: Boxwood & Chicago, IL: Quadrangle.

For more complete bibliographies, see Ashford (1985) and Ashford et al. (1993).

Biographies and Reviews of Richardson's Work

Ashford, Oliver M (1985) *Prophet – or Professor? The Life and Work of Lewis F Richardson*. Bristol: Hilger.

Boulding, Kenneth (1962) *Conflict and Defense. A General Theory*. New York: Harper and Row.

Gold, Ernest (1954) Lewis Fry Richardson. In: *Obituary Notices of Fellows of the Royal Society* 9: 217–235, http://rsbm.royalsocietypublishing.org/content/royobits/9/1/216.

Hess, George D (1995) An introduction to Lewis F Richardson and his mathematical theory of war and peace. *Conflict Management and Peace Science* 14(1): 77–113.

Hunt, Julian CR (1993) A general introduction to the life and work of Lewis F Richardson. In: Ashford et al., 2, 1–27.

Hunt, Julian CR (1998) Lewis F Richardson and his contributions to mathematics, meteorology, and peace research. *Annual Review of Fluid Mechanics* 30: xiii–xxxiv, https://www.annualreviews.org/doi/abs/10.1146/annurev.fluid.30.1.0.

Körner, Thomas William (1996) Ch. 8 A Quaker mathematician and Ch. 9 Richardson on War. In *The Pleasures of Counting*. Cambridge: Cambridge University Press, 159–193, 194–227.

Nicholson, Michael (1999) Lewis Fry Richardson and the study of the causes of war. *British Journal of Political Science* 29(3): 541–563.

Nowlan, Robert A (n.d.) Lewis F Richardson. In: A Chronicle of Mathematical People, http://209.151.25.133/pdfs/Richardson,%20Lewis%20Fry.pdf.

O'Connor, John J & Edmund F Robertson. The MacTutor History of Mathematics archive, entry for Lewis Fry Richardson, University of St Andrews, https://www-history.mcs.st-andrews.ac.uk/Biographies/Richardson.html.

Platzman, GW (1967) A retrospective view of Richardson's book on weather prediction. *Bulletin of the American Meteorological Society* 48(8): 514–550.

Poulton, EC (1993) The quantifying of mental events and sensations. In: Ashford et al. 2, 491–514.

Rapoport, Anatol (1957) Lewis F Richardson's mathematical theory of war. *Journal of Conflict Resolution* 1(3): 249–299. Also in: *General Systems Yearbook* 2: 55–90.

Rashevsky, Nicolas & Ernesto Trucco (1960) Preface. In: Lewis F Richardson: *Arms and Insecurity. A Mathematical Study of the Causes and Origins of War*. Pittsburg, PA: Boxwood, v–xi.

Richardson, Stephen A (1957) Lewis Fry Richardson (1881–1953): A personal biography. *Journal of Conflict Resolution* 1(3): 300–304.

Sutherland, Ian (1993) The causation of wars. In: Ashford et al., 32–42.

Wikipedia, Lewis Fry Richardson, https://en.wikipedia.org/wiki/Lewis_Fry_Richardson.

Wilkinson, David O (1980) *Deadly Quarrels. Lewis F Richardson and the Statistical Study of War*. Berkeley, CA: University of California Press.

Wright, Quincy & CC Lienau (1960) Editors' Introduction. In: Lewis F Richardson: *Statistics of Deadly Quarrels*. Pittsburgh, PA: Boxwood, v–xvii.

A more extensive list, including obituaries and reviews of his books, is found in Ashford (1985: 272–273).

Archival Sources

Paisley College of Technology, where Richardson taught for eleven years has a large collection of lecture notes, correspondence, and books from Richardson's personal collections with handwritten notes. The Richardson Institute, University of Lancaster, holds a collection of background material relating to his work in peace research. The Meteorological Office and numerous universities and other academic institutions have collections of his correspondence. For details, see Ashford (1985: 278–279).

About the Editor Nils Petter Gleditsch

Nils Petter Gleditsch (born 17 July 1942 in Sutton, Surrey, UK) is a Norwegian peace researcher and political scientist. He is Research Professor at the Peace Research Institute Oslo (PRIO). In 2009, Nils Petter Gleditsch was given the *Award for Outstanding Research* by the Research Council of Norway. In 1982 he was convicted (with Owen Wilkes) in Norway of a violation of the national security paragraphs of the penal code and given a suspended prison sentence for a publication on signals intelligence. After studies in philosophy and economics Gleditsch became mag.art. in sociology at the University of Oslo. In 1966–67 he read sociology, social psychology, and international relations at the University of Michigan. Since 1964, Gleditsch has worked at the Peace Research Institute Oslo (PRIO), first as a student, later as researcher. He was Director of PRIO in 1972 and 1977–78. From 2002 to 2008 he led the working group 'Environmental Factors of Civil War' at PRIO's Centre for the Study of Civil War, appointed as a Centre of Excellence by the Research Council of Norway. From 1993 to 2013 he was also been a part-time Professor of Political Science at NTNU, where he is now Professor Emeritus. Gleditsch was editor of Journal of Peace Research 1983–2010. He served as President for the International Studies Association (ISA) 2008–09. He is a member of the Royal Norwegian Society of Sciences and Letters (DKNVS) and the Norwegian Academy of Science and Letters (DNVA).

Among his books in English are: (coed., 1980): *Johan Galtung. A Bibliography of His Scholarly and Popular Writings 1951–80*; (with O Wilkes, 1987): *Loran-C and Omega. A Study of the Military Importance of Radio Navigation Aid*; (co-ed. with O Njølstad, 1990); *Arms Races – Technological and Political Dynamics*; (co-author with O Bjerkholt; Å Cappelen, 1994): *The Wages of Peace. Disarmament in a Small Industrialized Economy*; (co-ed. with O Bjerkholt; Å Cappelen; RP Smith; JP Dunne, 1996: *The Peace Dividend*; (co-ed. with L Brock; T Homer-Dixon; R Perelet;

© The Author(s) 2020

147

N. P. Gleditsch (ed.), *Lewis Fry Richardson: His Intellectual Legacy and Influence in the Social Sciences*, Pioneers in Arts, Humanities, Science, Engineering, Practice 27, https://doi.org/10.1007/978-3-030-31589-4

E Vlachos, 1997): *Conflict and the Environment*; (co-ed. with G Lindgren; N Mouhleb; S Smit; I de Soysa, 2000): *Making Peace Pay: A Bibliography on Disarmament and Conversion*; (co-ed. with P Diehl, 2001) *Environmental Conflict*; (co-ed. with G Schneider; K Barbieri 2003): *Globalization and Armed Conflict*; (co.ed. with G Schneider, 2013): *Assessing the Capitalist Peace*. He has also been guest-ed. and co-ed. of special journal issues including: *European Journal of International Relations*, 1(4): 405–574; *Political Geography* 26(6): 627–735; *International Studies Review* 12(1): 1–104; *International Interactions* 36(2): 107–213; *Conflict Management and Peace Science* 28(1): 5–85; *International Interactions* 38(4): 375–569; *Journal of Peace Research* 49(1): 1–267; *International Studies Perspectives* 13(3): 211–234; *International Studies Review* 15(3): 396–419.

Website at PRIO: http://www.prio.org/staff/npg.
Website at NTNU: http://www.ntnu.edu/employees/nilspg.
Website on the editor is at: http://afes-press-books.de/html/SpringerBriefs_PSP_Gleditsch.htm.

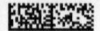